Energy Management and Operating Costs in Buildings

JOIN US ON THE INTERNET VIA WWW, GOPHER, FTP OR EMAIL:

WWW: http://www.thomson.com
GOPHER: gopher.thomson.com
FTP: ftp.thomson.com
EMAIL: findit@kiosk.thomson.com

A service of I(T)P

Other titles from E & FN Spon

Facilities Management: Theory and practice
Edited by Keith Alexander

Ventilation of Buildings
Hazim B. Awbi

Air Conditioning: A practical introduction
2nd edition
David V. Chadderton

Building Services Engineering
2nd edition
David V. Chadderton

Building Services Engineering Spreadsheets
David V. Chadderton

Spon's Mechanical and Electrical Services Price Book
Davis Langdon & Everest

Building Energy Management Systems
Geoff Levermore

Illustrated Encyclopedia of Building Services
David Kut

Heating and Water Services Design in Buildings
Keith Moss

Global Warming and the Built Environment
Edited by Robert Samuels and Deo K. Prasad

Site Management of Building Services Contractors
Jim Wild

Hot Water Details
Leslie Wooley and Phil Stronach

For more information about these and other titles please contact:
The Marketing Department, E & FN Spon, 2–6 Boundary Row, London,
SE1 8HN. Tel: 0171 865 0066

Energy Management and Operating Costs in Buildings

Keith J Moss I. ENG., ACIBSE

Visiting lecturer in Building Services Engineering to
The City of Bath College and the University of Bath

E & FN SPON
An Imprint of Chapman & Hall

London · Weinheim · New York · Tokyo · Melbourne · Madras

Published by E & FN Spon, an imprint of Chapman & Hall, 2–6 Boundary Row, London SE1 8HN, UK

Chapman & Hall, 2–6 Boundary Row, London SE1 8HN, UK

Chapman & Hall, GmbH, Pappelallee 3, 69469 Weinheim, Germany

Chapman & Hall USA, 115 Fifth Avenue, New York, NY 10003, USA

Chapman & Hall Japan, ITP-Japan, Kyowa Building, 3F, 2-2-1 Hirakawacho, Chiyoda-ku, Tokyo 102, Japan

Chapman & Hall Australia, 102 Dodds Street, South Melbourne, Victoria 3205, Australia

Chapman & Hall India, R. Seshadri, 32 Second Main Road, CIT East, Madras 600 035, India

First edition 1997

© 1997 Keith J. Moss

Typeset in 10½/12pt Sabon by WestKey Ltd, Falmouth, Cornwall

Printed in Great Britain by the Alden Press, Oxford

ISBN 0 419 21770 3

♾ Printed on acid-free text paper, manufactured in accordance with ANSI/NISO Z39.48-192 and ANSI/NISO Z39-1984 (Permanence of Paper).

Contents

Preface

Energy Management and Operating Costs in Buildings is a textbook for undergraduate courses in building services engineering, building engineering, energy engineering, the BTEC continuing education diploma and the higher national certificate in building services engineering. Since the management of energy in buildings requires an understanding of the behaviour of the building structure as well as the services in response to changes in outdoor climate, part of the text at least is appropriate to students studying building, construction management and building surveying.

Since the oil crisis of the 1970s and later the Rio Earth Summit 1992 the western world has been obligated to take responsibility for its profligate and indifferent use of the world's fossil fuel resources. Forty-five per cent of the UK's energy consumption is taken up by its use in the provision of space heating, space cooling, lighting, communications, hot water supply and cooking.

Industry in the last ten years has made considerable progress in cutting the consumption of fossil fuel and electricity, much of which is derived from fossil fuel, by changing its manufacturing processes and installing energy efficient plant.

Some of the larger organizations now have sophisticated energy conservation programmes for the services in their building stock. It is necessary now to encourage those organizations which have no energy management policy to join the fossil fuel stewardship campaign. Negotiating a competitive fuel tariff with a fuel supplier should be considered only as the first step in the management of energy consumption on the site or campus.

This book is one of only a few at present which addresses the methodologies of estimating annual energy consumption, undertaking energy audits and monitoring and targeting energy consumption. It discusses the background of each chapter and this is followed up with the appropriate underpinning knowledge and examples and case studies. References are made to source organizations, journals and articles which are pertinent to this important subject.

Acknowledgements

Grateful thanks are due to a number of part time students at Bath College, some of whom are working in the field of energy management, who have contributed to the preparation of the book and in teasing out many of the finer points in presentation.

I am also grateful to the following organizations, some of whom have allowed me to reproduce material relevant to the subject, much of which is recent in the form of articles in the professional journals listed elsewhere:

The Building Research Energy Conservation Support Unit
The Building Services Research and Information Association
The Chartered Institution of Building Services Engineers
The Energy Efficiency Office (Department of Energy)
The Energy Technology Support Unit
The Heating and Ventilating Contractors Association
The Meteorological Office

Those who have assisted me in the preparation of the book have not checked the arithmetic or solutions to examples and case studies. The author holds responsibility for these.

Introduction

Energy management and operating costs in buildings is a branch of the discipline of building services engineering. It is not essential that students wishing to specialize in energy management should have a prerequisite knowledge and experience of building services. In practice the background of many individuals who currently have a responsibility as energy managers varies widely. Many would agree however that underpinning knowledge about the services within the building can be of great help and continuing professional development in this area for those without it is invaluable. This book however is written in such a way as to develop basic skills in energy management regardless of prior professional training and where it is helpful, in Chapter 7 for example, the reader is directed to other reading matter.

Each chapter of the book is set out with the nomenclature used, an introduction, worked examples and case studies and data and text appropriate to the topic, and concludes with a chapter closure which identifies the skills and competencies acquired. The level of mathematics needed to gain full benefit from the text is between GCSE and A level and is introduced in the solutions where necessary. Computer programs and spreadsheets have not been used in the presentation material since the purpose of the book is to develop the underpinning knowledge of the subject. It is left to the reader to investigate the ever-moving market for software in this field.

Energy management is a moving feast; it is still a relatively young discipline and much of what is known is derived from historical data in the form of fuel invoices equated with building type, level of services specifications and level of building specifications. It is therefore necessary for the energy manager to keep abreast with current developments, publications and practices. A bibliography is included for the reader's benefit. It also benefited the author in the preparation of this book.

The economics of space heating plants 1

Nomenclature

A	area (m^2)
AEC	annual energy consumption (J, etc., kWh)
d	temperature rise due to indoor heat gains (K)
DD	Degree Day(s)
dt	design temperature difference (K)
e	exponent, e = 2.7183
F_1, F_2	temperature ratios
f_r	thermal response factor
HWS	Hot water supply
k	constant
MDD	maximum Degree Days
N	number of air changes/hour
Q_g	indoor heat gains (kW)
Q_p	plant energy output (kW)
S	number of days in the period under review
SDD	Standard Degree Day(s)
t_{ai}	indoor air temperature
t_{ao}	outdoor air temperature
t_b, B	base temperature in °C
t_c	dry resultant temperature
t_m	mean outdoor air temperature
t_n	minimum daily outdoor temperature
t_x	maximum daily outdoor temperature
U	thermal transmittance coefficient (W/m^2 K)
V	volume of space (m^3)
Y	thermal admittance (W/m^2 K)

1.1 Introduction

The use of primary fuels in industrialized countries has been the subject of national interest only since the oil crisis of the early 1970s. Much work has been done in the UK and elsewhere to reduce the consumption of energy derived from fossil fuels since that time, particularly in the manufacturing industries. Energy used in building services is estimated at 45% of national primary energy consumption in the UK, of which space heating, hot water supply and auxiliary power is estimated at 32%.

The building services industry therefore has a mandate here to design systems which conserve energy, to provide accurate forecasts of energy consumption, to promote energy conservation, to undertake energy audits and to monitor and target the consumption and future use of energy in buildings.

Clearly the building services engineer may only be responsible for one or two of these tasks but he or she should be able to contribute to all of them in a professional manner if called upon to do so. Energy managers on the other hand have a responsibility for all of these tasks.

Technical innovation and breakthrough in recent years has meant that boiler plant and associated equipment is more efficient, bringing the benefits of lower consumption of primary energy and less harmful releases of the products of combustion into the atmosphere.

It is likely that as this trend continues, with the increasing public awareness of issues which have a direct bearing on the well-being of our Earth, more emphasis will need to be placed upon the design of space heating plants in the world's temperate climates and hence upon costs in use. This is due also to the release of harmful products into the atmosphere during manufacturing processes as well as during the activities of building construction and as a result of living and working in the buildings we create for ourselves. It does now appear that controlling the release of carbon dioxide and pollutants may have a more significant impact on the building services industry than that resulting from the diminishing reserves of fossil fuel.

1.2 The economics

The economics of space heating plant should therefore be more relevant now than ever before. The economic evaluation of building engineering systems includes:

- capital costs;
- costs in use;
- life cycle costs;
- investment appraisal.

Investment appraisal is considered in Chapter 8.

Life cycle costs relate to the estimated life of plant and systems and to the following factors: reliability, maintainability and safety. The features given in Table 1.1 will influence the life cycle costs.

Investment for the replacement or refurbishment of plant and equipment may form part of the life cycle costs.

The life of the plant and equipment may be less than the life of the distribution pipework and radiators, for example. It may therefore be prudent to consider plant and equipment separately from distribution when accounting for replacement and refurbishment. See also Chapter 8.

Table 1.1 Factors affecting life cycle costs

Reliability	Maintainability	Safety
Degree of standardization	Spares availability	Statutory
Spares availability	Degree of complexity	Local specific
Standby requirement	Maintenance intervals	
	Ease of maintenance	
	Down time	
	Life	

Capital costs include:

- design fees;
- fees for supervising the installation of the services;
- material and labour costs of the installation;
- commissioning costs;
- costs for supplying the utilities of gas, water and electricity;
- builder's work and attendance costs.

Costs in use (operational costs) include:

- fuel;
- auxiliary power for boilers, pumps, fans, temperature controls etc.;
- preventive maintenance;
- corrective maintenance;
- insurance.

On large sites the client may consider buying in the supervision, operation and maintenance of the entire space heating plant or plants together with the management of its operational costs under an out-sourcing agreement referred to as Contract Energy Management. Purchase of new plant can also form part of an outsourcing agreement.

FUEL CONSUMPTION

The estimation for fuel consumption for space heating depends upon:

- plant energy output;
- seasonal efficiency of boiler plant and systems;
- duration of occupied period;
- mode of plant operation;
- thermal inertia of the building;
- internal heat sources;
- Degree Days appropriate to the season and the locality.

PLANT ENERGY OUTPUT Q_p.

This is calculated from the design heat loss for the building and is determined from the following formulae. (It is important to note here that in the calculation of weekly, monthly or annual fuel consumption, design heat loss is used. This is equivalent to plant energy output Q_p regardless of whether the plant in question operates continuously or intermittently.) Thus:

$$\text{Design heat loss: } Q_p = (\Sigma(UA)F_1 + 0.33NVF_2)(t_c - t_{ao}) \text{ W} \qquad (1.1)$$

where F_1 and F_2 are temperature ratios involving the indoor design comfort temperature t_c, the indoor environmental temperature t_{ei} and indoor air temperature t_{ai}.

The temperature ratios are dependent upon the proportions of radiant to convective heat emanating from the space heating appliances. They can be determined from data in the CIBSE Guide [1] or from formulae in another publication in the series [2]. Clearly the heat flow path resulting from a system of natural draught convectors will start at the indoor air point and proceed to the dry resultant, environmental and finally mean radiant point. In the case of a system of high temperature radiant strip or tube the resulting heat flow path commences at the mean radiant point. Thus the temperature ratios are affected by the type of heating system proposed for the building and in turn the building design heat loss is also affected. This matter is discussed at length in another publication in the series [2].

If, however, the building is thermally insulated to current standards and has low infiltration rates, which is to say that it is well sealed from ingress of outdoor air, design heat loss can be determined in the traditional manner from:

$$(\Sigma(UA) + 0.33NV)(t_{ai} - t_{ao}) \quad \text{W}$$

with only limited loss in accuracy.

SEASONAL EFFICIENCY OF BOILER PLANT AND SYSTEMS

Boiler manufacturers' quote efficiencies of their products under test conditions at full load. The efficiency of modern boilers ranges from 85% to 98% compared with boilers manufactured over ten years ago where the range was in the region of 70% to 80%. There is therefore a strong argument to replace boiler plant which is over ten years old.

A well-designed conventional boiler should maintain its efficiency to a turn down ratio of 30% of full load. Thereafter its efficiency will fall away and this is one of the reasons why modular boilers are recommended for plant energy outputs above 100 kW.

Condensing boilers, on the other hand, have efficiencies in excess of 90% in appropriate applications. They can be used with conventional boilers to improve the overall efficiency of the plant, particularly at

Table 1.2 Suggested seasonal efficiencies for plant and system

Type of system	Seasonal efficiency
Continuous space heating	
condenser/conventional boilers, compensated system	85%
fully controlled oil/gas fired radiator/convector system	70%
fully controlled oil/gas fired radiator/convector system with multiple modular boilers and sequence control	75%
Intermittent space heating	
condenser/conventional boilers, compensated system	80%
fully controlled oil/gas fired radiator/convector system	65%
fully controlled oil/gas fired radiator/convector system with multiple modular boilers and sequence control	70%
Domestic hot water	
gas/oil fired boiler and central storage	60%
direct gas/oil fired instantaneous water heaters	75%

low load and low temperature. Seasonal efficiency differs from efficiency under test conditions since it accounts for variations in load throughout the heating season.

In the case of the generation of hot water supply, central storage heated indirectly from the boiler plant will clearly have a lower seasonal efficiency than direct fired instantaneous HWS generation. Table 1.2 provides a guide to seasonal efficiencies.

DURATION OF THE OCCUPIED PERIOD

This varies with the use to which the building is put. For buildings intermittently occupied, like offices for example, space heating plant can be shut down at night and over weekends. It is normal to provide frost protection, where the plant will be triggered if the indoor temperature falls below a datum of around 12°C.

It is important that the building envelope on the warm side of the thermal insulation layer does not fall too low in temperature to avoid a long preheat period in addition to its original purpose of providing protection from water systems freezing. Consideration needs to be given to how the cleaning staff are accommodated when time scheduling the plant. For buildings intermittently occupied a correction is made to the annual SDD total. This matter is addressed in Chapter 3.

Buildings continuously occupied are continuously heated during the heating season. Space heating plant can have the facility of night setback which will cause a dip in the thermal capacity of the building envelope as well as the designed drop in indoor temperature.

EFFECTS OF MODE OF PLANT OPERATION ON FUEL CONSUMPTION

Plant energy output Q_p is dependent upon the mode of plant operation during the heating season and the thermal inertia of the building envelope. For continuously heated buildings having a high thermal inertia (this corresponds to a high thermal response factor f_r), plant energy output will not have to respond to the full downward swings in outdoor temperature during the heating season unless it is sustained over many days. This is due to the thermal damping effect of the building envelope which is assisted by the massive constituents of the internal walls and floors. For space heating plant operated intermittently, plant energy output does not vary much with variations in the thermal inertia of the building envelope.

For buildings with a low thermal response factor and intermittently occupied, the preheat period is relatively short and so is the cool down period after plant shut down which will take place just before the building is vacated.

On the other hand buildings having a high thermal inertia will require a longer preheat before occupation but the plant can be shut down some time before the building is vacated without loss in thermal comfort.

The effect on the length of the preheat period will, however, be significant after a weekend shut down for a building envelope having a high thermal inertia.

Location of the thermal insulation in the building envelope will also have an effect upon the length of the preheat period. Insulation located on the inside surface of external walls and ceilings of an envelope which otherwise has a high thermal inertia will effectively change the characteristics of the envelope to one of low thermal inertia: that is to say there will be a rapid response to the switching of the heating plant. This is because there is little thermal mass on the warm side of the insulation and in consequence the preheat and cool down periods will be short. Refer to Figure 1.1.

It should be emphasized here that the design heat load used for estimating annual energy consumption must not include a plant margin overload capacity or boosted plant output. Allowances for the daily, weekly and annual occupation times for buildings heated intermittently are made by correcting the SDD for the locality. This is investigated in Chapter 3.

EFFECTS OF MODE OF PLANT OPERATION ON THE THERMAL CHARACTERISTICS OF THE BUILDING ENVELOPE

As space heating plant becomes more intermittent in operation in response to occupancy patterns, heat flow **absorbed by** the building

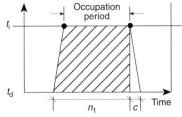

Intermittent heating insulation on inside surface

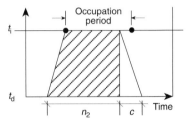

Intermittent heating insulation at the mid-point

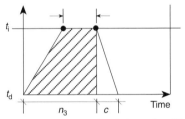

Highly intermittent heating insulation at the mid-point

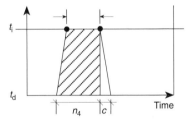

Highly intermittent heating insulation on the inside surface

Figure 1.1 The effect on plant operation of thermal insulation location in the building envelope: n_1 n_2; n_3 n_4 = plant operating hours including preheat; c = cooling time; t_d = datum temperature; t_i = indoor temperature.

envelope takes on significance from heat flow **transmitted through** the envelope. This is because the external structure of the building is cold and will absorb larger quantities of heat energy than that required for transmission through the structure.

The *CIBSE Guide* publishes tables of thermal transmittance coefficients (U values) and admittance (Y absorption values). For a cavity wall with 50 mm of glass fibre slab on the warm side of the air cavity, the U value is 0.46 W/m^2 K and the Y value is 3.7 W/m^2 K [3]. This clearly demonstrates that the absorption of heat energy by the cold external wall is many times more significant than the transmission of heat energy through it – eight times more significant in fact.

Once that part of the external wall on the warm side of the thermal insulation has risen from its datum temperature (t_d) following a shut down period to its optimum temperature, at the end of the preheat period, occupation of the building can take place without complaints of discomfort which otherwise would result from excessive body heat loss by heat radiation to the cold internal surfaces of the external envelope. Highly intermittent occupation patterns therefore require plant energy output to be determined from the admittance Y in place of transmittance U. Clearly the result is a space heating plant having a considerable overload capacity and the requirement for a long preheat period. The net result in terms of annual energy consumption is that similar buildings with intermittent and highly intermittent occupation patterns will have comparable consumption levels. Two examples of buildings having intermittent and highly intermittent occupancy patterns respectively are offices and some churches.

1.3 Internal heat sources

Internal heat gains Q_g are not normally considered when determining the plant energy output unless they are continuous. The automatic temperature controls on the space heating plant should account for internal heat gains. They should also account for solar heat gains, particularly through glazing, which occur on some winter days when the sun is at low altitude.

Internal heat gains are, however, considered when estimating weekly, monthly or annual fuel consumption. It is therefore necessary to access heat output data from various sources such as the human body, artificial lighting, computers, photocopiers, cooking and laundry equipment, display units, etc. The *CIBSE Guide* [4] and *BSRIA Rules of Thumb* [5] can be used as resource material although it is important not to overestimate the internal heat gains [6]. This allows the determination of the control temperature or base temperature t_b which is defined as the outdoor temperature above which no heating is required since the design indoor temperature will be maintained by the internal heat gains.

The temperature rise d due to the effect of internal heat gains is determined from:

$$d = Q_g/(\Sigma(UA) + 0.33NV) \text{ K}.$$

Alternatively

$$d = Q_g/(Q_p/dt) \text{ K}$$

where dt is the design temperature difference between indoors and outdoors.

Thus

base temperature $t_b = t_c - d$ °C.

Internal heat sources directly affect the determination of Degree Days. Standard Degree Days are calculated from a base temperature of 15.5°C.

The first matter which must be addressed here is why we use Degree Days to estimate projected fuel consumption and to undertake plant efficiency checks.

If the base temperature for a building is 15.5°C and prevailing outdoor temperature during occupancy is 20°C for half of a month and 12°C for the other half, the average outdoor temperature is 16°C. Since this is close to the base temperature inside the building it might

1.4 Standard Degree Days

Example 1.1

The plant energy output for a building is 210 kW for indoor and outdoor design temperatures of 20°C and −3°C, respectively. From the data, calculate the temperature rise due to indoor heat gains and hence determine the base temperature.

Data: 200 people at 140 W each, heating effect from artificial lighting 10 W/m², heating effect from office machinery evenly spread is 5 kW.
Total floor area 2300 m².

Solution
From the data

indoor heat gains $Q_g = 200 \times 140 + 10 \times 2300 + 5000 = 56\ 000$ W
indoor temperature rise $d = 56/(210/23) = 6.13$ K
base temperature indoors $t_b = t_c - d = 20 - 6.13 = 13.87$°C
base temperature = 13.87°C.

Clearly, since base temperature varies from the standard value of 15.5°C, a correction must be applied to the weekly, monthly and annual SDD. See Table 1.4.

The base temperature is also known as the balance temperature, which is related to continuous performance monitoring and is discussed in Chapter 10.

seem to indicate that the building requires no heating during the month in question. However, the fact of the matter is that for half the time the building is occupied it will require heating. So the SDD has been devised to account for the time the plant is in operation to trigger the fact that fuel is being used by the space heating plant. The average annual nine-month heating SDDs are given in Table 1.5 below for the 20 years to May 1979. Further SDD data is given in Appendix 1.

The determination of Degree Days is based upon the daily maximum t_x and minimum t_n outdoor temperatures and the temperature rise d resulting from internal heat gains. Monthly and annual Standard Degree Days (SDD) are published regularly in the journal of the Energy Efficiency Office [7] and are given for a base temperature of 15.5°C. There is also a *Fuel Efficiency Booklet* on heating Degree Days published by the Energy Efficiency Office [8] in which there is a map of the UK identifying the locations from which weather data is collected for the calculation of SDDs. The map shows Degree Day isopleths. Isopleths are isograms which are lines drawn on the map connecting points having equal numbers of Degree Days. There is also much other useful information in this booklet relating to SDDs and their applications. Standard Degree Days for seventeen locations in the UK and for a 20-year period to 1979 are given in the *CIBSE Guide* and reproduced in Table 1.5.

Monthly SDD totals for a specified locality vary for corresponding months each year due to variations in the seasons, particularly in temperate climates. Annual SDD totals therefore vary from year to year and also average annual SDD totals vary for each 10- or 20-year period considered. You will therefore find the monthly and annual SDD totals varying for a locality, depending on what period the SDD data is for. If this is a cause for concern the period taken for the purposes of calculating the annual fuel estimate should be stated in the submission to the client.

The Degree Day is dependent upon outdoor climate and indoor heat sources other than the space heating plant. Degree Days do not account for solar heat gain in winter or wind speed which adds a chill factor to the maximum and minimum outdoor temperatures if they are taken in a typically ventilated and screened local meteorological station.

Within these limitations the annual, monthly or weekly totals of Degree Days therefore provide the means of comparing over different periods and in different geographical locations the variations in load sustained by heating plants. They can also be used to check the consistency or otherwise of the performance of a heating plant on a weekly, monthly or annual basis.

Note: the apparent loss in efficiency in solutions to Examples 1.3 and 1.4 may be due to a variety of causes: the approach of a servicing

Example 1.2
A group heating scheme to a housing estate operates in the Thames Valley area. Determine the increase in energy consumption for a similar scheme projected for a location in Yorkshire.

Data: take the actual annual number of SDDs as 2012 for the Thames Valley and 2298 for Yorkshire.
    It is clearly apparent that the severity of the climate directly affects the number of Degree Days recorded for a locality.

Solution
The SDDs given in this example are taken from a 20-year period to 1971:

Estimate of the energy increase $= (2298 - 2012)/2012 = 0.14 = 14\%$.

Example 1.3
A building uses 150 litres of heating oil during a winter month having 380 DD.
    The consumption in the previous month having the same number of DDs was 144 litres. Calculate the apparent loss on plant efficiency.

Solution
Apparent loss in efficiency $= 100 \times (150 - 144)/144 = 4.16\%$.

Example 1.4
A building energy manager logs a fuel consumption of 178 litres of oil during a winter month having 341 DD. According to the records of a previous month having 351 DD the log shows an oil consumption of 170 litres. Determine the apparent effect, if any, on plant efficiency.

Solution
Current fuel consumption $= 178/341 = 0.522$ litres/DD.
Previous fuel consumption $= 170/351 = 0.484$ litres/DD.
Apparent loss in efficiency $= 100 \times (0.522 - 0.484)/0.484 = 7.85\%$.

contract, fuel delivery, making an error in recording fuel consumption, adjustment to the time scheduling of a heating circuit, windows left open during redecoration, lowering of combustion efficiency in the boiler

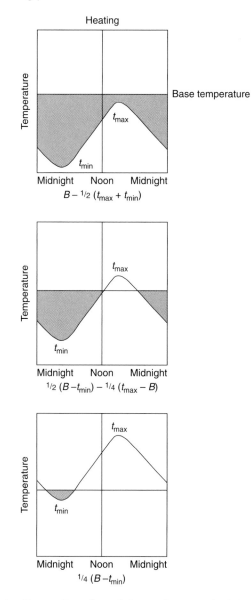

Figure 1.2 Estimation of heating Degree Days from daily maximum and minimum outdoor temperatures. Source: *Energy Efficiency Booklet No. 7.*

plant, thermal insulation breakdown in an external duct, etc. One of the tasks of the energy manager is to find the cause.

1.5 Calculation of Standard Degree Days

Reference should be made to *Fuel Efficiency Booklet No. 7* [8]. The actual number of Degree Days for a given location is assessed using maximum t_x and minimum t_n daily outdoor temperatures rather than the arithmetic mean daily temperature t_m outdoors. Refer to Figure 1.2

which illustrates the British Gas method of calculating Standard Degree Days.

Determination of Standard Degree Days adopting the empirical British Gas method

For t_x above 15.5°C but by a lesser amount than t_n is below

$$SDD/day = 0.5(15.5 - t_n) - 0.25(t_x - 15.5).$$

For t_x above 15.5°C by a greater amount than t_n below

$$SDD/day = 0.25(15.5 - t_n).$$

Clearly when t_x and t_n are both below 15.5°C

$$SDD/day = 15.5 - 0.5(t_x + t_n).$$

Thus

$$SDD/day = (15.5 - t_m).$$

The maximum possible number of Degree Days annually is determined from:

$$MDD = S(t_c - d - t_{ao}) = S(t_c - t_b)$$

where S is the number of days in the heating season, taken as 273, and t_{ao} is the outdoor design temperature (°C).

Note:

- The number of SDDs/day increases with the severity of the outdoor climate. The number of SDDs from Monday to Friday in this sample is low relative to the maximum possible number for five days which occurs when the outdoor design temperature remains consistently at −1°C.

Example 1.5

Determine the number of SDDs in the sample week taken from data recorded in a specified locality . What is the maximum number of DDs for the same period given an outdoor design temperature of −1°C?

Data:

Day	t_x	t_n
Monday	16°C	8°C
Tuesday	18°C	14°C
Wednesday	17°C	8°C
Thursday	13°C	6°C
Friday	10°C	2°C

Example 1.5 *continued*

Solution
The solution is given in Table 1.3.

Table 1.3 Solution to Example 1.5

Day	Calculation	SDD/day
Monday	$0.5(15.5 - 8) - 0.25(16 - 15.5)$	3.625
Tuesday	$0.25(15.5 - 14)$	0.375
Wednesday	$0.5(15.5 - 8) - 0.25(17 - 15.5)$	3.375
Thursday	$15.5 - 19/2$	6.000
Friday	$15.5 - 12/2$	9.500

From Table 1.3 the actual number of SDDs can be obtained by addition:

Actual number of SDD from Monday to Friday is 22.875.

The maximum number of SDDs is obtained by substituting the outdoor design temperature t_{ao} of $-1°C$ for the mean daily temperature t_{md}:

Maximum number of SDD $= 5(15.5 + 1) = 82.500$.

- The maximum possible number of Degree Days is calculated here so that a comparison with the recorded total SDD can be made.
- Little energy is required from the heating plant during the week and then only on Thursday and Friday.
- An approximate method for the calculation of SDDs for each day may be undertaken by subtracting the 24-hour mean daily temperature t_{md}, that is the mean of t_x and t_n, from the base temperature. The total for the five days considered in Example 1.5 on this alternative basis comes to 21.5 SDDs in comparison with 19.375 SDDs. Do you agree? This approximate method is used to determine monthly Degree Day totals from the adoption of Hitchin's formula. See Example 1.7 below.

1.6 Non-Standard Degree Days

Of course many buildings will have a base temperature different from the standard of 15.5°C (Example 1.1) which forms the basis of the published SDD data. In such cases this requires a correction to be made to the published SDD totals before they can be used in the determination of estimated energy use, comparisons of energy use, or efficiency checks. The *CIBSE Guide* has a table of correction factors reproduced here in Table 1.4.

Table 1.4 Corrections for base
temperatures other than 15.5°C

Base temperature (°C)	DD/SDD
10	0.33
12	0.57
14	0.82
15	0.94
15.5	1.0
16	1.06
17	1.18
18	1.3

The table for annual SDDs is published by the Meteorological Office and updated on a monthly basis [7]. Table 1.5 is taken from the CIBSE Guide and is the nine-month annual average for the 20-year period to May 1979 from 1 September to 31 May.

Further nine-month annual SDD data is to be found in Appendix 1.

From the map of isopleths in the *Fuel Efficiency Booklet No. 7*, it is apparent that there are habitable locations in which the nine-month annual SDDs exceed 2800. You will see from Table 1.5 that the highest SDD recorded for the period is 2617 for Aberdeen. It is important therefore to check the location of the project in hand with the map of isopleths.

The annual SDDs vary depending upon which 20-year period is

Table 1.5 Nine-month annual 20-year average of heating SDDs

Degree Day areas	Location	SDD
Thames Valley	Heathrow	2034
South Eastern	Gatwick	2275
Southern	Bournemouth	2130
South Western	Plymouth	1840
Severn Valley	Bristol	2109
Midlands	Birmingham	2357
West Pennines	Manchester	2233
North Western	Carlisle	2355
Borders	Boulmer	2464
North Eastern	Leeming	2354
East Pennines	Finningley	2243
East Anglia	Honington	2304
West Scotland	Glasgow	2399
East Scotland	Leuchars	2496
North East Scotland	Aberdeen	2617
Wales	Aberport	2094
Northern Ireland	Belfast	2330

Example 1.6
The building in Example 1.1 has a temperature rise due to indoor heat gains d of 6.13 K resulting in a base temperature t_b of 13.87°C. Determine the annual number of Degree Days if it is located in Wales.

Solution
From Table 1.5 SDD = 2094; from Table 1.4 DD/SDD = 0.57 when t_b = 12°C and 0.82 when t_b = 14°C. Adopting a linear interpolation for when t_b = 13.87°C, DD/SDD = 0.80375. Do you agree?

Thus the corrected annual Degree Days for this building assuming continuous heating will be DD = 2094 × 0.80375 = 1683.

Annual Degree Days = 1683.

considered since the UK climate is inconsistent, although it does vary within a recognized band.

Monthly and annual SDD data are based upon a heating season from 1 September to 31 May, which is 39 weeks or 273 days. In practice, heating plant is not started until it is sufficiently cold, which may extend well into October, in which case the annual SDD should be adjusted accordingly.

For practical purposes the annual number of SDDs, therefore, will vary depending upon the severity of the climate and the length of the heating season. Examples 1.6 and 1.7 are based upon the 20-year average annual SDD having 273 days in the heating season.

Further corrections to SDDs are required for intermittent heating. See Chapter 3.

1.7 Research into Degree Days

A more reliable method of determining Degree Days to any base is to use an empirical formula developed by Hitchin [8]:

$$\text{Average DD/day} = (t_b - t_m)/(1 - \exp[-k(t_b - t_m)])$$

where t_m is the mean outdoor air temperature in the month obtained from daily maximum and minimum temperatures and k, which varies slightly with the location, has been determined from 20 years of weather data from 1952 to 1971 and has a mean value of 0.71. The average Degree Days/day obtained from Hitchin's formula is then multiplied by the number of days in the month to obtain the monthly DD for a locality.

There has been much investigation done into the determination of Degree Days both for heating and cooling. One researcher defines Degree Days succinctly as the accumulated temperature which in turn is defined as the integrated excess or deficiency of external air temperature with reference to a fixed datum, namely base temperature [9].

A past Technical Secretary of CIBSE submitted a paper on how Degree Days are calculated and introduced a method for computing the 24-hour mean outdoor air temperature for each month of the year [10]. In the same paper quoted in [9] the author discusses more accurate Degree Day calculations obtained from hourly outdoor temperature records. It shows that errors occur when correcting the weekly, monthly and annual SDDs using Table 1.4 at low base temperatures.

A paper presented in *Building Services, the CIBSE Journal*, details an alternative method of determining AEC without the use of SDD [11].

Another paper presented in the same journal argues the case for Mirror Days [12].

There now follows an example adopting Hitchin's formula.

Example 1.7
Given that the base temperature for a building is calculated from indoor heat gains to be 12°C and the average outdoor temperature for the month of November was 6°C, determine the average DD/day and hence the DD for that particular month.

Solution
Substituting the data into Hitchin's empirical formula:

$$
\begin{aligned}
\text{Average DD/day} &= (12-6)/(1-e^{-0.71(12-6)}) \\
&= 6/(1-2.7183^{-4.26}) \\
&= 6/0.9859 \\
&= 6.09.
\end{aligned}
$$

The DD for the month of November will be $6.09 \times 30 = 183$ for a base temperature of 12°C.

CONCLUSION

Hitchin's formula can be used to generate monthly Degree Day totals for use in monitoring and targeting. See Chapter 10.

Daily maximum and minimum outdoor temperatures are recorded in meteorological stations in 18 locations around the UK for the determination of SDDs, but the determination does not account for the effects in the building of low altitude solar radiation during the winter season. Neither are the effects of wind chill factors accounted for. The

1.8 Limitations of the Degree Day method of estimating annual energy consumption

subjective difference between a dry cold climate and a damp cold climate is not accounted for either [13].

It may be appropriate to introduce a diversity factor to offset the effects of solar heat gains in a building in which there is local control of the heating appliances. The diversity factor could be applied as a correction to the annual SDD selected from Table 1.5. You may want to consider this factor when looking at Chapters 2 and 3.

1.9 Chapter closure   This chapter has focused upon the factors which might affect the cost in use of space heating plants in temperate climates and has introduced the concept of Standard Degree Days as a means by which energy costs might be measured and checks in the efficiency of plant and systems can be verified on a regular basis. Limitations on the accuracy of the application of SDDs have been identified so that value judgements can be made by the practising engineer or energy manager. References have been given for further study.

Estimating energy consumption – space heating 2

A	area (m^2)
AEC	annual energy consumption (J, etc., kWh)
AED	annual energy demand (J, etc., kWh)
AFC	annual fuel consumption (litres, tonnes, m^3)
AFc	annual fuel cost
CPV	cumulative present value
CV	calorific value (MJ/litre, etc.)
DD	Degree Day(s)
dt	design indoor/outdoor temperature difference (K)
EH	Equivalent Hours of operation at full load
MDD	Maximum Degree Days
N	number of air changes/hour
n	length of term in years
P_1, P_2	proportions of heat flow
Q	design heat loss (kW)
Q_p	plant energy output (kW)
Q_v	heat loss due to infiltration of outdoor air (W)
R	thermal resistance of material (m^2 K/W)
r	interest rate
R_{si}	inside thermal resistance (m^2 K/W)
R_{so}	outside thermal resistance (m^2 K/W)
R_t	total thermal resistance (m^2 K/W)
SDD	Standard Degree Day(s)
t_{ai}	indoor air temperature
t_{ao}	outdoor air temperature
t_c	dry resultant temperature
U	thermal transmittance coefficient (W/m^2 K)
V	volume (m^3)
WF	weather factor

2.1 Introduction

This chapter focuses upon the projected energy costs of space heating a continuously heated building which has yet to be commissioned and energy savings resulting from the application of thermal insulation to existing continuously heated buildings. Projected energy costs may be a requirement in the client's specification for the building services together with the determination of estimates of annual carbon dioxide and oxides of nitrogen emissions from the boiler plant. The request may be made at the appraisal stage when the project is still in its infancy in which case recourse to data from the *BSRIA Rules of Thumb* [1] and boiler manufacturers may be necessary.

2.2 Estimating procedures for continuously heated buildings

Examples of continuously heated buildings are hospitals, clinics, residential homes, nursing homes, workshops and factories on three-shift operation. There are four ways of calculating the annual energy demand (AED) for a projected building, namely:

- direct use of the annual number of Degree Days with the product of the thermal transmittance coefficient and associated area of the relevant part of the building envelope;
- direct use of the annual Degree Days with the design heat load per degree Q_p/K;
- adoption of load/weather factor WF with design heat load Q_p;
- use of the equivalent hours of plant operation at full load with the design heat load Q_p.

It will be shown that the determination of energy demand by the traditional method using the weather factor is now discounted.

It has to be stressed at the outset that these methodologies can only be estimates because of a number of unknown factors such as:

- overtime working;
- arrangements for cleaners;
- level of plant and system maintenance;
- level of supervision for day to day plant operation;
- level of occupants awareness of energy conservation.

These and other factors are not likely to be known at the design and specification stage of a project and will influence the AED estimate. This has been demonstrated in the technical press many times in recent years where building owners/occupiers have undertaken energy conservation measures by addressing some or all of the factors listed above, making substantial low/nil cost savings simply by good housekeeping.

One commences the estimation of AED for a projected plant with the design heat loss Q_p which includes the structural heat loss and that due to infiltration of outdoor air. It may be calculated in the traditional manner from indoor and outdoor design air temperatures or from dry resultant (comfort) temperature, outdoor air temperature and the temperature ratios F_1 and F_2. See equation (1.1).

If the building envelope is well insulated the simpler formula for design heat loss Q

$$Q = (\Sigma(UA) + 0.33\,NV)(t_{ai} - t_{ao}) \quad W,$$

may be adopted without introducing more than a 5% error. One method of estimating AED is to use the annual SDD for the locality, converting the days to seconds. Thus, adopting equation (1.1) in W/K:

$$AED = SDD \times 24 \times 3600 \times Q_p/(t_c - t_{ao}) \quad J. \qquad (2.1)$$

Checking the units:

Degree Days $\times$ hr/day $\times$ sec/hr $\times$ J/sK = J.

Alternatively, if Q_p is in kW, $AED = SDD \times 24 \times Q_p/(t_c - t_{ao})$ kWh. $\qquad (2.2)$

Checking the units:

Degree Days $\times$ hr/day $\times$ kW/K = kWh.

Example 2.1
A projected building having a design heat loss of 180 kW is to be located in the Midlands. Determine the annual energy demand for continuous heating during the heating season from 1 September to 31 May. Indoor and outdoor design temperatures are 19°C and −3°C, respectively.

Solution
Adopting equation (2.1) and SDDs from Table 1.5

$$AED = 2357 \times 24 \times 3600 \times 180\,000/(19 + 3) = 1670\,GJ.$$

Adopting equation (2.2)

$$AED = 2357 \times 24 \times 180/22 = 462\,829\,kWh.$$

Note: (i) the seasonal efficiency is not accounted for in AED; (ii) you can see from these solutions that 1 kWh = 3.6 MJ.

The example which follows considers the potential energy savings resulting from roof insulation.

Example 2.2
The roof of an industrial building located in East Anglia is to be insulated such that its thermal transmittance coefficient is reduced from 3.13 to 0.4 W/m^2 K. If the roof area is 400 m^2 estimate the annual saving in energy demand for a continuously heated building using SDDs.

Solution
Here equation (2.1) can be adapted by replacing Q_p/K with UA

$$AED = SDD \times 24 \times 3600 \times UA \quad J. \tag{2.3}$$

Checking the units:

$$\text{Degree Days} \times hr/day \times sec/hr \times J/sm^2 \, K \times m^2 = J.$$

You will note that indoor/outdoor design temperatures are not given in the question and in fact are not needed for the solution. However, the adoption of SDDs implies the use of the base temperature of 15.5°C, the average outdoor temperature over the winter season and continuous heating. Corrected DDs can be used instead here for different base temperatures and for intermittent heating. Refer to Chapter 3. The SDD for East Anglia is taken from Table 1.5.
  Substituting data before insulation:

$$AED = 2304 \times 24 \times 3600 \times 3.13 \times 400 = 249 \, GJ.$$

After insulation:

$$AED = 2304 \times 24 \times 3600 \times 0.4 \times 400 = 31 \, GJ.$$

The annual estimate of energy savings is $249 - 31 = 218 \, GJ.$

Note: AED does not account for seasonal efficiency.
  These solutions can be readily converted to litres of heating oil, tonnes of coal or m^3 of natural gas or left in kWh and corrected for seasonal plant efficiency. Table 2.1 lists the calorific values of different fuels.
  Account must be taken of the seasonal efficiency of the space heating system when calculating AEC, AFC or AFc, for example:

AEC = AED/seasonal efficiency
AFC = AED/(CV × seasonal efficiency)
AFc = (AED × cost per unit)/(CV per unit × seasonal efficiency).

Table 2.1 Calorific values of different fuels

Fuel	Calorific value
Gas oil 35s	37.8 MJ/litre
Light oil 250s	40.5 MJ/litre
Medium oil 1000s	40.9 MJ/litre
Butane	28.3 MJ/litre
Propane	25.3 MJ/litre
Natural gas	38.7 MJ/m^3
Coal (average)	27.4 MJ/kg
Electricity	3.6 MJ/kWh

Example 2.3

A building located in the North East of Scotland is to be continuously heated and has a heat loss of 220 kW at design conditions of 20°C and −5°C. Determine the storage volume for medium grade oil required on site for a three week period based upon (a) the annual SDD data and (b) a continuous outdoor temperature of −5°C. Seasonal efficiency can be taken as 65%.

Solution

Annual fuel consumption can be determined from:

$$\text{AFC} = \text{AED}/(\text{CV} \times \text{efficiency}) \text{ litres of oil.} \tag{2.4}$$

Checking the units:

$$\text{MJ}/((\text{MJ/litre}) \times \text{efficiency}) = \text{litres.}$$

It is important to reconcile the units in the formula here by substituting the design heat loss in MW. Combining equations (2.1) and (2.4):

$$\text{AFC} = (\text{SDD} \times 24 \times 3600 \times Q_p/K)/(\text{CV} \times \text{efficiency}).$$

SDD for North East Scotland is taken from Table 1.5 and

$$\text{AFC} = (2617 \times 24 \times 3600 \times 0.220/25)/(40.9 \times 0.65)$$
$$= 74\,845 \text{ litres of oil.}$$

For a three-week period based upon SDDs, the estimated volume to be stored on site will be $74\,845 \times 3/39 = 5.76\,\text{m}^3$.

(a) Volume of oil requiring storage $= 5.76\,\text{m}^3$.

This assumes a 39-week heating season. The estimated volume to be stored on site based upon a continuous outdoor temperature of −5°C will be:

$$21 \text{ days} \times 24 \times 3600 \times \text{design heat loss}/(\text{CV} \times \text{efficiency}).$$

Thus storage volume $= (21 \times 24 \times 3600 \times 0.22)/(40.9 \times 0.65)$
$$= 15\,015 \text{ litres.}$$

(b) Volume of oil requiring storage $= 15\,\text{m}^3$.

Note the absence of design temperature difference in the solution here since we have 21 days not 21 Degree Days.

You can clearly see the effect of three weeks of severe weather upon the consumption of fuel oil here. Another more realistic estimate could be made from the average three weeks of highest recorded SDD for North East Scotland over a 20-year period. If you can locate weekly SDDs for this area [2] what is the storage volume and how does it compare with the above solutions? Does it lie between the two?

With the exception of space heating by electricity, seasonal efficiency for space heating plant usually lies between 60% and 75%. See Table 1.2.

Example 2.3 estimates the site storage volume of fuel oil.

The following case study considers a financial appraisal of adding roof insulation.

Case study 2.1

The roof of a workshop in continuous operation is 'flat' and consists of 12 mm stone chips, three layers of felt on 20 mm of shuttering ply over 100 mm wood joists with 1 mm of polythene and 6 mm of plasterboard on the underside. It is intended to line the underside of the plasterboard with 25 mm of polyurethane board.

(a) Given that the roof dimensions are 20 m × 15 m determine the annual fuel cost estimate before and after refurbishment.
(b) If the installation cost is £2500 determine:
   (i) the simple payback period;
   (ii) the payback period using a discount rate of 4%.

Data: adopt the SDD for the Thames Valley, fuel for space heating is light grade oil at a cost of 22p per litre, seasonal efficiency is taken as 65%.

SOLUTION

Part (a)
The first part of the solution requires a knowledge of how the thermal transmittance coefficient or U value is determined. Since it is a joisted roof there is differential heat flow, namely that through the joists U_j and that through the air space between them U_a. See Figure 2.1. This calls for the adoption of the formula for the non-standard thermal transmittance coefficient U_n where $U_n = U_j P_1 + U_a P_2$ and P_1 and P_2 are the proportions of heat flow

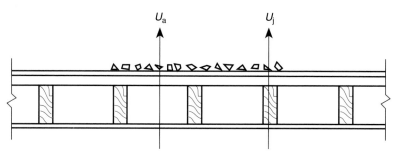

Figure 2.1   Roof detail for case study 2.1.

through the joists and through the air space between, taken as 30% and 70%, respectively.

The thermal conductivities of the constituent parts of the roof are taken from data in the *CIBSE Guide* [3] and the total thermal resistance R_t, before refurbishment, through the joisted portion of the roof is:

$$R_t = R_{so} + R_{stonechips} + R_{polythene} + R_{ply} + R_{joist} + R_{plasterboard} + R_{si}$$
$$= 0.04 + 0.0125 + 0.018 + 0.1429 + 0.7143 + 0.0375 + 0.1$$
$$= 1.0652 \, m^2 \, K/W$$

from which $U_j = 0.939 \, W/m^2 K$.

The total thermal resistance R_t through the air space between the joists before refurbishment will be:

$$R_t = 0.04 + 0.0125 + 0.018 + 0.1429 + 0.16 + 0.0375 + 0.1$$
$$= 0.5109 \, m^2 \, K/W,$$

from which $U_a = 1.96 \, W/m^2 \, K$.

After refurbishment the thermal resistance of the polyurethane insulation must be added to each calculation, from which $U_j = 0.484$ and $U_a = 0.660 \, W/m^2 \, K$. The non-standard thermal transmittance coefficients U_n can now be determined thus:

before refurbishment $U_n = 0.939 \times 0.3 + 1.96 \times 0.7$
$$= 1.6537 \, W/m^2 \, K$$
after refurbishment $U_n = 0.484 \times 0.3 + 0.66 \times 0.7$
$$= 0.6072 \, W/m^2 \, K.$$

We are now in a position to determine the annual energy cost of heat flow through the roof of the workshop before and after refurbishment. Combining equations (2.3) and (2.4) and including the cost of oil per litre:

before refurbishment AFc = $(U_n \times A \times SDD \times 24 \times 3600$
$$\times \, cost)/(CV \times efficiency).$$

The units of the quantities are:

$(kW/m^2 \, K) \times m^2 \times Degree \, Days \times (hrs/day) \times (sec/hr \times cost/litre)/$
$(kJ/litre)$.

Note that care must be taken to reconcile the units here with U_n in $kW/m^2 \, K$ and CV in kJ/litre. The heating SDD for the Thames Valley is taken from Table 1.5.

(a) Before the improvement

AFc = $(0.0016537 \times 20 \times 15 \times 2034 \times 24 \times 3600 \times 0.22)/$
    (40500×0.65).
    = £728.61.

(b) After the improvement

$$AFc = (0.0006072 \times 20 \times 15 \times 2034 \times 24 \times 3600 \times 0.22)/$$
$$(40\ 500 \times 0.65)$$
$$= £267.53.$$

There is a significant difference, as you can see, between the estimated costs before and after refurbishment.

Part (b)

The second part of the question refers to both simple and discounted payback periods. The payback period is of importance since the services engineer or energy manager will be required to produce evidence that the energy saving measure is cost effective. This will depend upon the attitude that the client or senior management has towards seeing a return on capital spent. Short payback periods are unfortunately the present trend partly due to current philosophy in the market place. This attitude may be forced to change by the effects of dwindling resources of primary energy and the greenhouse effect, allowing periods of payback to be extended over a number of years. If there are no costs in the use of the energy-saving measure, simple payback is the cost of the measure divided by the annual savings in fuel costs thus:

(i) Simple payback = (cost − annual maintenance cost)/
annual savings
$$= £2500/(£727 - £267) = 5.4 \text{ years.}$$

This means that after 5.4 years the building owner will be able to reap the benefit of the saving in fuel costs of £460 per annum, the cost of the improvement having been paid for by the annual savings in that payback period. It is assumed here that the work will be done when the workshop is not in use, otherwise loss in production must be included in the cost and this would extend the period of payback. Simple payback assumes that at the end of the payback period relative costs will be the same as they were at the beginning. Discounted payback attempts to allow for interest rates and is discussed in Chapter 8. The formula which applies here is that for cumulative present value, CPV, and

$$CPV = ((1 - (1 + r)^{-n})/r.$$

Statistical tables are published giving values of CPV and n for ascending values of r and are included in Appendix 6. Furthermore CPV = simple payback = 5.4; $r = 4\%$ and n is the number of years of payback. Thus

$$5.4 = (1 - (1.04)^{-n})/0.04$$

and

$$0.216 = 1 - (1.04)^{-n};$$

therefore

$(1.04)^{-n} = 0.784$

and

$(1.04)^{n} = 1.2755.$

Thus:

$n \times \log 1.04 = \log 1.2755$

and

$n = 0.1057/0.0170;$

therefore

$n = 6.2$ years.

(ii) Payback period discounted at 4% = 6.2 years.

The higher the discount rate the longer will be the period of payback. For more details of financial appraisal refer to Chapter 8.

There now follows an example of the effects of reducing the natural infiltration rate.

Example 2.4

A continually heated workshop measuring $18 \times 12 \times 4$ m high and located in Northern Ireland is to have draught inhibiting doors fitted and it is estimated that this will reduce the average heat loss due to infiltration from two air changes per hour to 0.75. Determine the annual savings estimate of medium grade fuel oil used in the space heating plant. Seasonal efficiency may be taken as 65%.

Solution

Clearly it is not a straightforward matter to estimate the reduction in infiltration of outdoor air resulting from the introduction of the new doors. The Building Services Research and Information Association however can undertake on-site determination on a before and after basis. Working with the information in the question and combining equations (2.1) and (2.4) and substituting Q_v/K, which is $0.33\,NV$, for Q_p/K, in equation (2.1), annual oil consumption due to infiltration is:

$\text{AFC} = (0.33\,NV \times \text{SDD} \times 24 \times 3600)/(1000 \times 0.65 \times \text{CV}).$

Check the terms and the units here; the SDD is taken from Table 1.5, the CV is taken from Table 2.1.

Units of terms = $(\text{kW/K}) \times \text{Degree Days} \times (\text{hr/day}) \times (\text{sec/hr})/$
$(\text{kJ/litre}) = \text{litres}.$

Consumption before improvement:

$$AFC = ((0.33 \times 2 \times 18 \times 12 \times 4) \times 2330 \times 24 \times 3600)/$$
$$(1000 \times 0.65 \times 40\,900)$$
$$= 4319\,litres.$$

Consumption after improvement:

$$AFC = ((0.33 \times 0.75 \times 864) \times 2330 \times 24 \times 3600)/$$
$$(1000 \times 0.65 \times 40\,900)$$
$$= 1619 \text{ litres.}$$

The estimate of the annual saving in fuel oil will be $4319 - 1619 = 2700$ litres.

2.3 Adoption of Equivalent Hours and the weather/load factor

THE WEATHER FACTOR

This is expressed as the ratio of annual SDD for a locality and the maximum number of DD annually; thus

$$WF = SDD/MDD.$$

For a plant operating continuously over the standard winter of 273 days:

$$AFC = (\text{design load in kW} \times 3600 \times 24 \times 273 \times WF)/(CV \times \text{efficiency}).$$

$$\text{Units of terms} = ((kJ/s) \times (sec/hr) \times (hr/day) \times days)/(kJ/tonne) = tonnes.$$

Example 2.5

A continuously heated building is to be located in the Borders and has a design heat load of 180 kW. Using the system of Standard Degree Days calculate the estimated fuel consumption annually by adopting the weather factor. This estimate should then be compared with the adoption of Equivalent Hours at full load and the adoption of Standard Degree Days.

Data: the fuel is coal, seasonal efficiency is 65%, indoor design temperature is 20°C and outdoor design temperature is −3°C.

Solution

SDD = 2464, from Table 1.5.
MDD = 273(15.5 + 3) = 5051.

Thus the weather factor is:

WF = 2464/5051 = 0.488.
$$AFC = (\text{design load} \times 3600 \times 24 \times 273 \times WF)/(CV \times 0.65 \text{ tonnes}).$$

The units of the terms are:

$(MW \times (sec/hr) \times (hr/day) \times days)/(MJ/tonne)$.
$AFC = (0.180 \times 3600 \times 24 \times 273 \times 0.488)/(27\,400 \times 0.65)$
$\qquad = 116.3\,tonnes$.

Note the conversion of the design heat load from kW to MW to reconcile the units.

Note also that the indoor design temperature has not been used in the solution and the base temperature has been taken as 15.5°C in the calculation of the WF.

EQUIVALENT HOURS OF OPERATION AT FULL LOAD, EH

An alternative approach to the estimate of annual energy consumption was first introduced in the *CIBSE Guide* in 1970 [4]. Here

$$AED = SDD \times 24 \times 3600 \times design\ heat\ load/dt$$

where dt is the design temperature difference (K). Units of terms = (Degree Days $\times$ (hr/day) $\times$ (sec/hr) $\times$ (kJ/s))/ K = kJ.

Rearranging the formula:

$$AED = design\ heat\ load\ kW \times 3600 \times (24\,SDD/dt)\quad kJ \qquad (2.5)$$

where the Equivalent Hours of operation at full load is:

$$EH = 24\,SDD/dt\ \ hours.$$

Thus:

$$EH = Standard\ annual\ degree\ hours/K.$$

EH therefore represents the actual running hours of the heating plant at full load, based upon the annual SDD for the locality. From equation (2.4):

$$AFC = AED/(CV \times efficiency).$$

Note the similarity between equations (2.1) and (2.5) which will in fact give the same results.

There is a clear distinction between the weather factor calculation of AFC and AFC determined from Equivalent Hours of plant operation at full load. Consider the formula using the weather factor:

$$AED = design\ load\ kW \times 3600 \times 24 \times 273 \times WF\ \ kJ.$$

This may be expanded into the form:

$$AED = design\ load \times 3600 \times 24 \times 273 \times SDD/MDD.$$

Then

$$AED = design\ load \times 3600 \times 24 \times 273 \times SDD/273(t_c - d - t_{ao}).$$

This reduces to:

$$AED = \text{design load } kW \times 3600 \times 24 \times SDD/(15.5 - t_{ao}) \quad kJ.$$

If this is compared with equations (2.1) and (2.5) the change lies in the temperature difference used in each formula where use of the WF involves base temperature and the use of EH, for example, involves indoor design temperature. Clearly AED and AFC calculated from equations (2.1) and (2.5) will yield lower annual estimates of energy demand and fuel consumption.

Consider Example 2.5; the annual fuel consumption estimate for the building located in the Borders, calculated by adopting the formula for equivalent hours of operation, equation (2.5), will be:

$$AED = 0.180 \times 3600 \times 24 \times 2464/(20 + 3) = 166\ 609 \text{ MJ}$$

and

$$AFC = AED/(CV \times \text{efficiency}),$$

so

$$AFC \quad = 166\ 609/(27\ 400 \times 0.65) = 93.5 \text{ tonnes.}$$

This compares with the estimate of 116.3 tonnes of coal annually, based on use of the weather factor. This is a decrease of almost 20% in the annual consumption of coal.

It is apparent that the discrepancy occurs because the adoption of Equivalent Hours of plant operation at full load and employment of equation (2.1), which adopts SDDs directly, do not take into account the temperature rise due to indoor heat gains.

DIRECT USE OF STANDARD DEGREE DAYS

Consider again Example 2.5. If the formulae (2.1) and (2.4) for annual fuel consumption are adopted for the solution to Example 2.5, a further comparison can be made.

$$AFC = (SDD \times 24 \times 3600 \times Q_p/dt)/(CV \times \text{efficiency}).$$

Substituting:

$$AFC = (2464 \times 24 \times 3600 \times 0.18/23)/(27\ 400 \times 0.65)$$
$$= 93.5 \text{ tonnes.}$$

This compares with the solution adopting equivalent hours of plant operation at full load EH.

Example 2.5 analyses the difference between use of the weather factor and the Equivalent Hours at full load for continuously heated buildings. The conclusion which follows from these comparisons is that the adoption of the load/weather factor is likely to lead to an overestimate of energy consumption. It was the original method adopted in the UK and historically it has been found to lead to overestimates.

There is another matter which, if taken into account, could have the effect of further increasing the discrepancy when the WF is used. The calculation of the annual maximum number of Degree Days was originally based upon an outdoor temperature at a continuous value of $-1°C$ regardless of the value of the outdoor temperature used in any design calculations. The outdoor temperature was adopted on the assumption that it would be difficult to comprehend a more severe winter than that anywhere in the UK. On this basis:

$$MDD = 273(15.5 + 1) = 4505.$$

The weather factor in Example 2.5 therefore would be $WF = 2464/4505 = 0.547$ and not 0.488, which results from an outdoor design temperature of $-3°C$. The resulting fuel consumption annually will be $AFC = 130$ tonnes. Do you agree?

In the light of current practice annual fuel estimates should be determined directly from SDDs or the use of Equivalent Hours of plant operation at full load, EH, in preference to the adoption of the weather/load factor.

2.4 The preferred model for estimating annual fuel consumption

You can clearly see from Example 2.6 the effect that the annual total of Standard Degree Days has on the number of hours the plant is operating at full load. This of course reflects the effect of the local winter climate and the fact that the outdoor temperature in the heating season is rarely as low as the outdoor design temperature for long periods of time.

2.5 Equivalent Hours and maximum hours at full load

It is most important when submitting a report of annual energy costs or savings to management or a client that it contains statements qualifying the estimate:

2.6 Qualifying remarks

- There are a number of factors relating to the accuracy of the Degree Day method for the calculation of costs. See Section 1.7.
- In the determination of the design heat loss it is assumed that natural infiltration of outdoor air occurs simultaneously in all the rooms in the building. In fact it will only take place on the windward side of the building. Thus the design heat loss attributable to natural infiltration is about 50% of the total at any one time.
- It is left to you to identify further qualifying remarks which could be included in the report. Have a look at Chapters 3 and 4.

Example 2.6
Compare the Equivalent Hours of plant operation at full load with the maximum number of hours at full load for the continuously heated building in Example 2.5.

Solution
Equivalent Hours of plant operation at full load is:

$$EH = 24\,SDD/dt_d$$
$$= 24 \times 2464/(20 + 3)$$
$$= 2571 \text{ hours.}$$

The maximum number of hours at full load will occur when the plant is operating continuously at design load over the heating season and

$$\text{Maximum annual hours} = 273 \times 24 = 6552 \text{ hours.}$$

2.7 Chapter closure

In this chapter we have investigated four methodologies for the determination of annual fuel estimates for continuously heated buildings. One of the methods, the first to be adopted in the UK, is shown historically to provide an overestimate and should therefore be discounted. For this reason it will not be adopted in Chapter 3.

You are now able to estimate the annual fuel consumption/cost for continuously heated buildings in the UK and elsewhere in temperate climates if Standard Degree Days are available. If an annual fuel cost estimate is required, it is important that the best tariff is obtained from the fuel supplier, not forgetting the standing charge which must be applied. This information should accompany the estimate when submission is made to the client along with the qualifying remarks.

Intermittent space heating 3

Nomenclature

AED   annual energy demand
AFC   annual fuel consumption
AFc   annual fuel cost
CV   calorific value
d temperature rise due to indoor heat gains (K)
DD   Degree Day(s)
dt design indoor/outdoor temperature difference (K)
E number of days occupation in the week
F number of weeks occupation in the year
HWS   hot water supply
MDD   maximum Degree Days
N number of occupants
Q_g heat gains indoors (kW)
Q_p plant energy output (kW)
SDD   Standard Degree Day(s)
t_{ao} outdoor air temperature
t_b base temperature in °C
t_c indoor comfort temperature in °C
Y consumption of hot water (litres/person)
z heat losses from hot water supply system

3.1 Introduction

Clearly not all buildings are heated continuously throughout the winter, primarily since they are not continuously occupied. As discussed in Chapter 1, this has an effect upon the building envelope and internal structure by causing it to absorb heat energy during the preheat period and beyond and to reject heat energy, preferably into the building, when the plant is shut down. The use of solid partitioning such as concrete blocks and floors assists in damping down this diurnal swing in the mean temperature of the building envelope and the temperature within the building. The location of the thermal insulation in the building envelope also dictates the magnitude and rate of swing in its mean temperature as plant is switched on and off.

These matters may be considered at the feasibility stage of a project and the location of the thermal insulation within the building envelope can be affected by the pattern of occupancy.

3.2 The estimation of annual energy demand

The estimation of annual energy demand, AED, for heated buildings, however, is based upon the design heat load without the addition of a plant margin, overload capacity or boosted plant output and this is true for intermittently heated as well as continuously heated buildings.

However, the thermal capacity and therefore the thermal response of intermittently heated buildings must be accounted for in view of the matters raised in Chapter 1 and the comments in the introduction to this chapter, in respect of building structure – light, medium or heavy weight – and the occupancy pattern.

Lightweight buildings are those of single storey factory type with little or no solid partitions, or top floors of multi-storey buildings when undivided. Mediumweight buildings are those with a single storey of masonry or concrete with solid partitions in concrete block. Heavyweight buildings are those of more than one storey and constructed from masonry or concrete with solid floors and some solid partitioning. Very heavy buildings are those of curtain walling, masonry or concrete, especially multi-storey, much subdivided within by solid partitions. Corrections are made to the weekly, monthly and nine month annual Standard Degree Day to account for these factors. These are taken from the *CIBSE Guide* [1] and given in Tables 3.1, 3.2 and 3.3.

Note that the location of the thermal insulation within the building envelope may change its thermal response; a heavyweight building can behave like a lightweight building if, during refurbishment, for example, thermal insulation is applied to the inside surface.

Table 3.1 Correction for the length of the occupied week

Occupied period	Light building	Heavy building
7 day	1.0	1.0
5 day	0.75	0.85

Table 3.2 Correction for length of occupied day

Occupied period	Light building	Heavy building
4 hours	0.68	0.96
8 hours	1.00	1.00
12 hours	1.25	1.02
16 hours	1.40	1.03

Table 3.3 Correction for response of building and plant

Type of heating	Light building	Medium building	Heavy building
Continuous	1.0	1.0	1.0
Intermittent – responsive plant	0.55	0.7	0.85
Intermittent – plant with long time lag	0.70	0.85	0.95

There are therefore a total of four potential corrections to the Standard Degree Day when temperature rise due to indoor heat gains is included, which is accounted for in Table 1.4. The corrected SDD then allows the estimation of AED for intermittently heated buildings.

Case study 3.1

Estimate the annual fuel consumption for a four storey office block located in Fife, East Scotland. The estimated heat gains from lighting and office equipment is 25 W/m^2.

3.3 The estimation of annual fuel consumption for an office

Data:

Design heat loss	115 kW
Floor area	360 m^2/floor
Indoor design temperature	20°C
Outdoor design temperature	−3°C
Occupancy pattern	five days a week, eight hours/day
Plant operation	responsive/intermittent
Fuel	medium grade oil
Seasonal efficiency	65%

SOLUTION

From Table 1.5 the annual SDD in Leuchars is 2496. Temperature rise due to heat gains indoors:

$$d = Q_g/(Q_p/dt)$$
$$= (360 \times 4 \times 0.025)/(115/(20 + 3))$$
$$= 7.2 \text{ K.}$$

Base temperature $t_b = t_c - d = 20 - 7.2 = 12.8°C$.

From Table 1.4 the correction factor for the temperature rise due to indoor heat gains is, by interpolation,

DD/SDD = 0.67.

From Table 3.1 for five days a week occupation correction = 0.85.
From Table 3.2 for eight hours a day occupation correction = 1.00.

From Table 3.3 for intermittent/responsive plant correction = 0.85. The corrected SDD is:

SDD = $2496 \times 0.67 \times 0.85 \times 1.00 \times 0.85 = 1208$.

From equation (2.3):

AED = design load $\times 3600 \times$ (24 DD/dt)
     = $115 \times 3600 \times (24 \times 1208/23) = 522$ GJ.

AFC = AED/(CV $\times$ efficiency) = $522\,000/(40.9 \times 0.65)$.

Note: CV of medium grade fuel oil from Table 2.1 is 40.9 MJ/litre, so

AFC = 19 635 litres.

A comparison with the adoption of the weather factor would now be appropriate as an addendum to case study 3.1.

SOLUTION

The maximum number of degree days for the standard year of 273 days will be:

MDD = $273(t_c - d - t_{ao}) = 273(20 - 7.2 + 3) = 4313$.

Corrected SDD for Fife to the base temperature of 12.8°C = $2496 \times 0.67 = 1672$.

Weather factor = SDD/MDD = $1672/4313 = 0.388$.

The occupancy is eight hours/day. Allowing for office cleaning and occasional extension of the eight hour period, assume an average daily plant run time of 12 hours.

Annual plant running hours = $(273/7) \times 5 \times 12 = 2340$ hours

Now

  AED = design load $\times$ running hours $\times 3600 \times$ weather factor   kJ.

The units are:

  (kJ/s) $\times$ h $\times$ s/h   kJ.

Substituting data:

AED = $115 \times 2340 \times 3600 \times 0.388 = 376$ GJ.

AEC = $376/0.65 = 578$ GJ.

AFC = $578 \times 1000/40.9 = 14\,132$ litres.

Summarizing case study 3.1
The adoption of the Weather Factor here shows a lower fuel consumption estimate than that obtained from the use of Equivalent Hours.

You will have noticed, however, that the Equivalent Hours calculation above was based upon a heavyweight building where the corrected annual Degree Days came to 1208. For a medium weight building the correction factors change and are: 0.8, 1.0 and 0.7 respectively. Thus the corrected annual degree days for Fife = $2496 \times 0.67 \times 0.8 \times 1.0 \times 0.7 = 936$.

Thus:

$$AED = 115 \times 3600 \times (24 \times 936/23) = 404 \text{ GJ}$$

and

$$AEC = 404/0.65 = 622 \text{ GJ}.$$

Then

$$AFC = 622 \times 1000/40.9 = 15\ 208 \text{ litres of oil}.$$

This is much closer to the estimate obtained by adopting the Weather Factor.

Clearly, the correction factors given in Tables 3.1, 3.2 and 3.3 for use with the Equivalent Hours calculation have a significant effect upon the annual energy demand and hence the AFC, and should be selected with care.

There now follows a more complex analysis of a day school.

Case study 3.2

A day school, largely single storey, is located in Belfast and has a design heat loss of 320 kW for an average indoor temperature of 19°C when the outdoor temperature is −1.5°C. It is equipped with indoor sports facilities which do not include a swimming pool. The two badminton courts and assembly hall are hired out for two nights a week and on Saturdays throughout the heating season to local community groups. The design heat loss for these areas is 22 kW and 27 kW respectively. Estimate the annual energy costs for the school and the estimated annual charge for heating the out-of-hours facilities.

Data:
Floor area of the school is 2900 m².
Fuel supply to the school is natural gas at a cost of 1.3p/kWh and an annual standing charge of £215.

3.4 Estimation of annual fuel consumption for a school

School holidays: Autumn half-term, one week; Christmas, two weeks; spring half-term, one week; Easter, two weeks; Spring bank holiday, one week.

During term times occupancy is five days a week, seven hours a day. During the holiday periods the plant is held on setback during weekdays at an indoor temperature of 15°C except for the office accommodation, the temperature controls of which can be set as required.

SOLUTION

Note: this solution will not account for fuel consumption for the catering facilities at the school. The nine-month SDD for Belfast is 2330 from Table 1.5. The correction factors from Tables 3.1, 3.2 and 3.3 are selected/interpolated as 0.75, 0.92 and 0.7, respectively. You will see that an intermittent plant with a long time lag has been chosen from Table 3.3 in the absence of appropriate information.

There is no reference to indoor heat gains in the question. One source of heat gain which is likely here across the whole school will be that due to lighting. Other sources, such as that from computer equipment, will be discounted since they are not likely to be evenly spread across the school.

A figure of 10 W/m^2 is a conservative estimate for a 500 lux level of illuminance from an enclosed surface-mounting fluorescent luminaire. Further information is contained in the *CIBSE Guide* [2] and Appendix 7.

The temperature rise due to the heat gain from lighting is

$$d = Q_g/(Q_p/dt)$$
$$= 2900 \times 10/(320\,000/20.5) = 1.86 \text{ K}.$$

If 2 K is taken as the temperature rise resulting from the lighting, the base temperature is $t_b = 19 - 2 = 17°C$ and from Table 1.4 the correction is DD/SDD = 1.18. Applying these corrections to the SDD total for Belfast:

corrected DD = $2330 \times 0.75 \times 0.92 \times 0.7 \times 1.18 = 1328$.

From equation (2.3):

AED = design load in kW $\times 3600 \times (24\text{DD}/dt)$ kJ
    = design load in kW $\times (24\text{DD}/dt)$ kWh
    = $320 \times 24 \times 1328/20.5 = 497\,514$ kWh.

The annual fuel cost is:

AFc = ((AED/efficiency) $\times$ cost/kWh) + annual standing charge.
    = $((497\,514/0.65) \times 1.3/100) + 215 = £10\,165$.

The following qualifying remarks need to be made otherwise the AFc estimate is open to serious criticism.

- A diversity factor can be applied to the AED for space heating to account for the penetration of low altitude solar radiation. Degree Days do not account for solar heat gains which will occur during the winter months.
- Holiday periods when the plant is set back during the day are not accounted for here, neither is the use of the offices during the holidays. It is assumed that the one will offset the other.
- Account must now be taken of the school facilities offered to the local community.

The application of Degree Days is not appropriate here. Evening occupation occurs when the outdoor temperature is likely to be dipping although the use of the assembly hall on Saturdays might be more typical of a Degree Day calculation. Thirty-nine weeks' use during the heating season at say three hours for each of two weekday evenings and 12 hours use on Saturdays will provide the total occupancy time:

$$39 \times 2 \times 3 + 39 \times 12 = 702 \text{ hours.}$$

This is the time that the space heating plant will be required to maintain the indoor design temperature. It does not represent the time the plant is operating at full load and therefore a correction factor must be applied to account for occasions when the outdoor temperature is above the value of $-1.5°C$ which was used for design purposes.

The maximum number of Degree Days annually for the nine-month heating season, based upon the design indoor and outdoor temperatures for the locality, will be:

$$\text{MDD} = 39 \times 7(19 + 1.5) = 5597.$$

The SDD for Belfast is 2330. The correction factor which will be applied here is $2330/5597 = 0.416$.

Note: this is **not** the weather or load factor discussed in Chapter 2 which was calculated using base temperature and not indoor design temperature. Thus, equivalent hours of plant operation at full load will be:

$$702 \times 0.416 = 292 \text{ hours.}$$

The design heat load is $22 + 27 = 49$ kW and AED is $49 \times 292 = 14\,308$ kWh. Thus:

$$\text{AFc} = (14308/0.65) \times 1.3/100 + (215 \times 14\,308/497\,514) = £293.$$

Estimated annual charge for heating the out of hours facilities = £293.

Note: another way of determining the charge is to look at the annual kWh totals: the total for the school is 497 514 kWh and that for the out-of-hours facilities is 14 308 kWh. This represents 2.88% of the total annual energy consumption and 2.88% of the annual fuel cost of £10 165 is £293.

What further qualifying remarks should be made relating to these AFcs?

3.6 The estimation of annual fuel consumption for a factory

The following case study relates to a factory operating a two-shift system.

Case study 3.3

A single storey factory situated on a trading estate near Birmingham having a two-shift system of production has a design heat loss of 240 kW when held at 17°C with an outdoor temperature of −3°C. It is heated with direct gas-fired unit heaters for six days per week, 50 weeks per year. The factory floor area is 2000 m^2. Estimate the annual fuel consumption for space heating.

SOLUTION

A production timetable of two shifts can be taken as two shifts of eight hours per day. The SDD for Birmingham, from Table 1.5, is 2357. From Tables 3.1, 3.2 and 3.3, the correction factors by interpolation, where appropriate, are: 0.875, 1.4 and 0.55 respectively. Do you agree?

The temperature rise indoors, d, due to heat gains is difficult to establish because there is no information relating to the presence or otherwise of machinery in the factory. However there will be lighting and at 500 lux a figure of 15 W/m^2 will be used assuming industrial trough fluorescent tubes (refer to the *CIBSE Guide* [2]).

Temperature rise:

$$d = Q_g/(Q_p/dt)$$
$$= 15 \times 2000/(240\ 000/(17 + 3))$$
$$= 2.5\ \text{K}.$$

Base temperature:

$$t_b = 17 - 2.5 = 14.5°C.$$

From Table 1.4,

DD/SDD = 0.88 by interpolation.

Applying these corrections to the nine-month annual Standard Degree Day total for Birmingham:

corrected DD = $2357 \times 0.875 \times 1.4 \times 0.55 \times 0.88 = 1397$.

From equation 2.3:

AED = design load $\times 3600 \times (24\ DD/dt)$
= $240 \times 3600 \times (24 \times 1397/20) = 1448$ GJ.

Annual gas consumption is AFC = AED/(CV × efficiency).
From Table 2.1 the CV of natural gas is 38.7 MJ/m^3, and since the heaters are direct fired units a seasonal efficiency of 75% will be taken. Thus:

AFC = $1448\,000/(38.7 \times 0.75) = 49\,888\ m^3$ of natural gas.

For the reader who has little experience of estimating annual energy consumption the final case study considers a domestic residence which hopefully will provide the vehicle for looking at the annual consumption of energy with which the reader is more familiar.

3.7 The estimation of annual fuel consumption for a house

Case study 3.4

A three-bedroom semi-detached family residence on two storeys having a floor area of 100 m^2 has a heat loss of 8 kW and is located in Cardiff. The indoor design temperature averaged through the house is taken as 19°C when the outdoor temperature is −1°C. Estimate the annual energy consumption for space heating and hot water supply. The house is of traditional build with cavity external walls and a ventilated wood-on-joist ground floor.

SOLUTION

For the purposes of calculation, and given the general construction characteristics of the house, it is assumed that the building is of medium weight construction. The temperature rise, d, due to indoor heat gains is limited to 3 K to ensure a comfortable thermal environment is attained at the expense of energy cost. Heating is provided initially for 16 hours a day, seven days a week during the heating season.

From Table 3.1 the correction for seven days occupation per week is 1.0. From Table 3.2 the correction for a length of day of 16 hours is 1.2. From Table 3.3 the correction for a responsive plant and intermittent heating is 0.7. From Table 1.4 the correction for a temperature rise due to indoor heat gains of 2 K which gives a base temperature of $t_b = 19 - 3 = 16$°C will be 1.06.

Applying these corrections to the annual SDD for Cardiff of 2094 the corrected annual DD for the locality will be:

corrected DD = $2094 \times 1.0 \times 1.2 \times 0.7 \times 1.06 = 1865$.

Adopting equation (2.3) to obtain the annual estimate for space heating the house:

$$
\begin{aligned}
AED &= \text{design load} \times 3600(24\ DD/dt) \\
&= 8 \times 3600 \times 24 \times 1865/(19 + 1) \\
&= 64.45\ \text{GJ}.
\end{aligned}
$$

The hot water supply for the house must now be considered. If the house is occupied for seven days per week, 50 weeks per year, then for a family of three the annual energy estimate for hot water supply can be determined from equation (4.1) in Chapter 4:

$$AED = 231 \times YN(1 + z/100)EF\ \text{kJ}.$$

From Table 4.1 the daily use of hot water Y for a medium rental dwelling is 30 litres/person and taking the heat losses as 10% we have:

$$AED = 231 \times 30 \times 3(1.1) \times 7 \times 50 = 8\ \text{GJ}.$$

Thus the annual total estimate is $65 + 8 = 73$ GJ.

If the boiler is fired by natural gas the current domestic rate is 1.433 pence/kWh and the standing charge is 9.02 pence per day with VAT at 8%. There is 0.0036 GJ in 1 kWh and taking a seasonal efficiency of 75%, the net annual cost estimate for space heating and hot water supply will be:

$$AFc = (((73/0.0036) \times 0.01433)/0.75) + 0.0902 \times 7 \times 50.$$

The units are kWh $\times$ ($£$ sterling/kWh) + $£$ sterling.

$$
\begin{aligned}
AFc &= 387.44 + 31.57 \\
&= £419\ \text{net} \\
&= £452\ \text{with VAT}.
\end{aligned}
$$

This is equivalent to £37.7 per month.

The provision of hot water supply represents $8/73 = 11\%$ of the annual energy estimate. This accounts for £50 of the annual fuel estimate and is therefore approximately £4.17 per month. If the external cavity walls are filled with insulation granules and the house is double glazed, resulting in a design heat loss of 6 kW, the annual energy cost estimate for space heating and hot water supply will be £355. This is a reduction of approximately 21% in the annual estimate. Do you agree with this?

The summary to case study 3.4 is given in Table 3.4 which includes further analysis.

SUMMARY OF CASE STUDY 3.4

Do you agree with the before and after insulation results in Table 3.4?
    You should now undertake another analysis in which the heating
system is on for three hours in the morning and six hours in the evening.
Adopt the thermally insulated model for the analysis assuming that the
house will retain some of the heat to provide a basic level of comfort
during the day between the hours of 09.00 when the boiler is turned
off and 16.00 when it is turned on again. The analysis for this time
scheduling arrangement is given in the last column of Table 3.4.
    Note that the HWS consumes an increasing proportion of the annual
energy requirement and that the effect of diversity has not been
accounted for. It is likely that if the heating appliances are individually
temperature controlled, solar heat gain in the winter season when the
sun is low in altitude and therefore penetrates well into the house will
aid in reducing energy consumption. Remember that Degree Days do
not account for the effects of solar radiation. It is also assumed that
the whole house is heated during boiler operating periods when it is
likely that the bedrooms, for example, will only be partially heated
during part of the operating period. A diversity factor is therefore
appropriate here for these two reasons. A diversity factor of 0.8 is
proposed. Would you agree? What effect will this have on the estimated
annual costs in Table 3.4?
    The Building Research Energy Conservation Support Unit
(BRECSU) has recently reviewed some modern ultra-low energy homes
[3] in which the total energy delivered is between 113 and 153 kWh/m².
You should compare this with the performance indicators in Table 3.4
which do not include electrical energy or energy for cooking.

Table 3.4 Summarizing the solution to case study 3.4

Item	House		
	Before insulation	After insulation	Further analysis
Heat loss	8 kW	6 kW	6 kW
Annual heating	65 GJ	48 GJ	40 GJ
Annual HWS	8 GJ	8 GJ	8 GJ
Annual total	73 GJ	56 GJ	48 GJ
% HWS	11%	14%	17%
Annual GJ/m²	0.73	0.56	0.48
Annual kWh/m²	203	156	133
Annual heating cost	£402	£305	£259
Annual HWS cost	£50	£50	£52
Annual total cost	£452	£355	£311
Monthly heating cost	£33.50	£25.42	£21.6
Monthly HWS cost	£4.17	£4.17	£4.3
Monthly total cost	£37.7	£29.6	£26
Reduction in annual cost	–	21%	31%

3.8 Further qualifying remarks

Each of these solutions must be supported with qualifying remarks. It would be quite unprofessional to submit the fuel estimate calculated in the solution to case study 3.3, for example, without any supporting statements to the client or to senior management. One would be leaving oneself open to criticism and the charge of naivety.

- Auxiliary power for the pumps, fans, controls, fuel burners, etc. is not accounted for in the estimate.
- There will be some machinery in the factory which will influence the heat gains.
- More accurate information relating to the luminaires and level of illumination would be useful.

Can you add to these qualifying remarks?

More general observations can also be made about the use of SDDs:

- The period over which the Annual Degree Days were recorded to obtain SDDs for the locality should be stated since they vary from one period to the next and from year to year.
- The place where the building is located may not compare climatically with the location from which the SDDs are calculated by the Meteorological Office.
- Degree Days do not account for the effects of solar heat gain, which can be significant through glazing due to the low altitude of the sun in the winter months.
- The chill factor resulting from wind speed is not accounted for in the determination of Degree Days.

Can you add to the qualifying remarks which have already been made? Note:

1. The selection and use of the correction factors must be done with care since they have a significant effect upon the calculation of the corrected DD and hence the energy/fuel estimate.
2. Case study 3.4 gives annual energy consumptions in GJ/m^2 of floor. This way of defining the energy consumption for space heating a building is termed the Performance Indicator and is considered in detail in Chapter 6.

3.9 Chapter closure

You now have the skills to estimate the annual energy and energy costs for three types of intermittently heated buildings in the UK and elsewhere in temperate climates if Standard Degree Days are available. The extension of these acquired skills to other building types is a matter of applying the principles learnt in this chapter. It requires reference to current data, access to historical data specific to the building type, if available, and the application of practical common sense. Remember the importance of the qualifying qemarks relating to the annual fuel estimate in your report.

Estimating the annual cost for the provision of hot water supply 4

Nomenclature

A	floor area (m^2)
AEC	annual energy consumption (GJ)
AED	annual energy demand (GJ)
AFC	annual fuel consumption
AFc	annual fuel cost (£ sterling)
CV	calorific value
E	number of days/week
F	number of weeks/year
HWS	hot water supply
N	number of occupants
P	probability of usage
q	HWS (W/m^2 of floor)
S	number of days under review
Y	hot water consumption (litres/person)
z	allowance for heat losses

4.1 Introduction

The provision of a hot water supply for use by the building occupants traditionally has been made by either the adoption of hot water heaters local to the point of use or by central generation via primary hot water connections from the space heating plant to a heat exchanger in the hot water supply cylinder or calorifier. Current technology now favours separation of space heating from hot water generation and a return to the generation of hot water supply directly in plant designed specifically for the purpose. This has the effect of increasing both the combustion efficiency and the thermal efficiency of hot water generation and hence reducing fuel consumption and emissions, with a consequent lowering of operating costs.

4.2 Factors to be considered

The following factors need to be addressed when consideration is given to the provision of a hot water supply since they will impinge upon the cost in use:

1. The number and type of fittings, for example: personal ablution, laundering, cooking, dishwashing.
2. The number of consumers served.
3. Simultaneous rates of flow which in all probability will require the adoption of the usage ratio P. There are very few applications for the provision of hot and cold water where all the draw off points on a system of hot water supply will be in use simultaneously. One exception to the rule is groups of showers in a school, clubhouse or factory. The determination of simultaneous rates of flow will influence the heat losses from the pipework and hence the cost in use.
4. Whether fittings are closely grouped or widely distributed. Clearly it is not economic to have long runs of distribution pipework from a central generating source serving fittings widely distributed around the building.
5. Nature of the water supply. This will have a bearing upon the possibility of scale build up and/or corrosion, the need for water treatment and the requirement for maintenance to maintain maximum efficiency of primary energy conversion.
6. Method of generation. If the hot water is generated and stored in a vessel there will be heat losses from the vessel as well as the circulating pipework which will contribute to the cost in use. The instantaneous generation of hot water on the other hand does not incur this cost and this contributes to the current preference for the provision of direct and instantaneous generation of hot water, although account is taken of heat losses in the circulating pipework. Account of the generating method adopted is taken in the choice of seasonal efficiency.
7. Storage/operating temperature. The current consensus is 65°C as a practical minimum to inhibit the growth of legionella spores. Excessive temperatures increase the effects of scale formation and corrosion. Storage/operating temperature clearly is directly related to costs in use. Factors 1, 2, 3 and 5 identified above also apply to the provision of cold water supply.

4.3 Hot water supply requirements and boiler power

The requirements for hot water vary in accordance with the type of occupation. One of the earlier editions of the *CIBSE Guide* [1] offers data in tabular form which has been adapted and reproduced here in Table 4.1. The column for boiler power is included but not used in the following examples.

The data in Table 4.1 is for guidance purposes in the absence of historical evidence.

Table 4.1 Daily hot water consumption and boiler power

Building	Daily water use at 65°C (litres/person)	Boiler power to 65°C (KW/person)
Boarding school	25	0.7
Day school/college dwellings:	5	0.1
low rental	25	0.5
medium rental	30	0.7
high rental	45	1.2
Factories	5	0.12
with showers	25	0.7
Hospitals:		
general	30	1.5
maternity	30	2.1
Offices	5	0.1
Sports centres	35	0.3
Hotels	40	1.3

If the daily demand for hot water is Y litres/person for the number of occupants, N, in the building having access to hot water, then the daily demand will be YN. If the water is heated from 10 to 65°C, the net daily heat requirement will be:

$$YN \times 4.2 \times (65 - 10) = 231YN \text{ kJ/day}.$$

If heat losses are $z\%$ the gross heat requirement will be:

$$231YN(1 + z/100) \text{ kJ/day}.$$

If the building is occupied for E days per week and F weeks per year:

$$AED = 231YN(1 + z/100)EF \text{ kJ} \tag{4.1}$$

and annual fuel consumption will be:

$$AFC = AED/(CV \times efficiency). \tag{4.2}$$

The units for this equation depend upon the units for the calorific value CV and therefore will be in litres, kilograms or cubic metres. If the annual energy consumption AEC is in kJ and CV is expressed in MJ, the annual fuel consumption is:

$$AFC = (231YN(1 + z/100)EF)/(1000 \times CV \times efficiency). \tag{4.3}$$

Equations (4.2) and (4.3) are therefore similar. Natural gas prices are now quoted in pence/kWh. The annual fuel cost AFc will therefore be:

$$AFc = (AED \text{ in kJ}/(1000 \times 3.6 \times efficiency)) \times cost/kWh \quad £ \tag{4.4}$$

Note the conversion 1 kWh = 3.6 MJ and the conversion of AED in kJ to MJ. Note also that the allowance for heat losses z in the circulating pipework and plant is taken as 10% in the following examples. You

4.4 Determination of the annual energy consumption estimate

should now consider equations (4.1), (4.2), (4.3) and (4.4) and satisfy yourself as to their differences and similarities.

Unlike space heating, the provision of a hot water supply will be required throughout the year during occupation periods. It is therefore considered as a base load. The subject of base loads is considered in detail in Chapter 10.

There now follows an example on the hot water requirements for a new boarding school.

Example 4.1

A new boarding school is to house 210 boarding pupils, 65 day pupils and 20 members of staff. Determine the annual estimate for cost in use related to the provision of a hot water supply. Assume indirect heating.

Solution

Clearly this is very limited information on which to base an estimate. The length of the terms are needed, for example, as are the extent of the sports, cooking, dishwashing and laundering facilities. An allowance will need to be made for occupation outside term time and the extent to which some of the school's facilities are rented out to other organizations. Not all the hot water will be generated from the same source; it is possible for example that some wash hand basins will be served by local electric hot water heaters.

If it is assumed that out of the 52 weeks of the year the school is in use for 32 weeks and the data in Table 4.1 is adopted, the following calculation can be made from equation (4.1). For the boarding pupils:

$$AED = 231 \times 25 \times 210(1.1) \times 7 \times 32 = 298.82 \text{ GJ}.$$

For day pupils, assuming they attend for 5.5 days a week:

$$AED = 231 \times 5 \times 65(1.1) \times 5.5 \times 32 = 14.53 \text{ GJ}.$$

For staff, assuming attendance is averaged out at 6 days per week for 38 weeks per year, equation (4.1) gives:

$$AED = 231 \times 5 \times 20(1.1) \times 6 \times 38 = 5.79 \text{ GJ}.$$

Therefore:

$$\text{total AED} = 319 \text{ GJ}.$$

If the fuel is natural gas:

$$CV = 38.7 \text{ MJ/m}^3$$

and the seasonal efficiency is 65%. The annual fuel consumption from equation (4.2) is:

$$AFC = AED \text{ in MJ/(CV} \times \text{efficiency)}$$
$$= 319\,000/(38.7 \times 0.65) = 12\,681 \text{ m}^3.$$

If the natural gas tariff is 1.5p/kWh, the annual fuel cost from equation (4.4) is:

$$AFc = (\text{energy in MJ}/(3.6 \times \text{efficiency})) \times \text{cost per kWh}$$
$$= (319\,000/(3.6 \times 0.65)) \times 1.5/100 = £2045.$$

Note the conversion from MJ to kWh where 1 kWh = 3.6 MJ. Note also that there will be a standing charge which must be added to the annual cost estimate.

When submitting an estimate of this kind it is important to include any qualifying remarks which help to provide the background to the estimate.

4.5 Qualifying remarks – boarding school

- It is necessary to negotiate the tariff with the fuel supplier. It depends upon the quantity of fuel consumed annually, the location of the project and the supplier. Fuel tariffs are considered in Chapter 9.
- This annual fuel estimate is based upon very few hard facts. It therefore needs qualification. You would be open to major criticism if the factors taken in preparing the estimate were not accounted for. Have a look at the qualifications listed in the solution to Example 4.3 and then make a list of qualifications suited to this solution.

The following example considers the hot water requirements for a factory.

Example 4.2
A factory operating on a two-shift system has a total of 120 occupants of which 15 are office staff, 70 are machine operatives and the remainder are manual staff. The office staff do a single nine-hour shift for five days a week and half a day on Saturday. There are two groups of employees who attend the machines and do the manual work, each group undertaking a ten-hour shift over seven days. All employees get the statutory holidays and two weeks a year paid leave. Estimate the annual fuel consumption from centralized direct oil fired heaters operating on light grade fuel oil.

Solution
For the purposes of this estimate the statutory holidays will be discounted. There are three grades of work here, namely office, machine and manual. It will be assumed that the manual workers will require a shower at the end of their shift. The annual number of working days EF for the office staff will be 50 weeks $\times$ 5.5 days

and the figure of 5 litres per person from Table 4.1 will be used. Now from equation (4.1):

$$AED = 231YN(1 + z/100)EF \quad kJ$$
$$= 231 \times 5 \times 15(1.1) \times 5.5 \times 50 = 5.241 \quad GJ.$$

The annual number of working days EF per shift will be 50 weeks $\times 7$ days and for the machine operators an estimate of ten litres per person per shift will be used in the absence of data. Do you agree with this estimate of hot water consumption for machine operators? Thus from equation (4.1):

$$AED = 231 \times 10 \times 70(1.1) \times 7 \times 50 = 62.25 \; GJ \text{ per shift.}$$

For two shifts of machine operators:

$$AED = 124.5 \; GJ.$$

For manual workers the figure of 25 litres per person per shift from Table 4.1 will be adopted. Thus from equation (4.1):

$$AED = 231 \times 25 \times 35(1.1) \times 7 \times 50 = 77.82 \; GJ \text{ per shift.}$$

For two shifts of manual workers:

$$AED = 155.64 \; GJ.$$

The total annual energy consumption is:

$$AEC = (5.241 + 124.5 + 155.64) = 285 \; GJ.$$

From Table 2.1 the calorific value for light fuel oil is $CV = 40.5$ MJ/litre. Adopting equation (4.2) the annual fuel consumption for the provision of hot water supply will be:

$$AFC = AED \text{ in } MJ/(CV \times \text{efficiency}).$$

The seasonal efficiency of direct fired oil heaters for the hot water supply like their gas fired counterparts will be relatively high if they are adequately maintained, a figure of 75% would be reasonable.

$$\text{Estimated } AFC = 285\,000/(40.5 \times 0.75) = 9383 \text{ litres of light fuel oil.}$$

This is a base load since it is required throughout the year and is not dependent upon the heating season. Qualifying remarks should support this estimate.

The next example considers the hot water requirements for a sports centre.

Example 4.3

A sports centre has the following facilities: international size swimming pool, toddlers' pool, five squash courts, two basket ball courts, weight training gym, aerobic gym, four games rooms, café and restaurant. The centre has five male and female shower rooms to support the sporting activities. It also has five male and female toilets. Estimate the annual consumption of natural gas associated with the provision of hot water for the centre.

Solution

There are a number of approaches to this solution. Clearly fuel consumption estimates based upon historical data from a similar sports centre would prove invaluable here. In the absence of this kind of data one approach is to estimate the occupancy in the various activities offered at the centre and apply a diversity factor.

Consumption rates for hot water are taken from Table 4.1 where possible. Other rates of consumption are estimated. The data is given in Table 4.2 to arrive at a daily total hot water consumption for the centre. The column headed 'Daily use' is an estimate of the number of times or sessions in the day that the facility is used.

Table 4.2 Data for the solution to Example 4.3

Activity areas	Occupancy estimate	Daily use	Hot water consumption (litres per person)	Total consumption
Main pool	100	6	35	21 000
Toddlers' pool	50	6	5	1500
Five squash courts	10	8	35	2800
Two basket ball courts	20	8	35	5600
Weight training gym	20	6	35	4200
Aerobic gym	30	8	35	8400
Four games rooms	40	6	5	1200
Café	20	8	1	160
Restaurant	20	4	5	400
Staff	20	2	5	200
Total persons	330/session	2020/day	Daily consumption	45 460

From the totals the maximum daily occupancy for the sports centre is estimated at 2020. This figure does not include spectators. The total consumption of hot water resulting from this occupancy is estimated as 45 460 litres.

It is unlikely that all activities will be at full capacity on any one day and a diversity factor of 40% will be adopted here. This reduces the daily provision of hot water to 18 184 litres.

The average consumption of hot water per person before application of the diversity factor will be 45 460/2020 = 22.5 litres/day.

For the purposes of this fuel estimate it will be assumed that the sports centre is open for 51 weeks in the year, seven days per week, and the seasonal efficiency of indirect gas fired plant is 65%. From equation (4.1):

$$AED = 231 \times 18\ 184 \times (1.1) \times 7 \times 51 = 1650 \text{ GJ}.$$

From equation (4.2):

$$AFC = 1\ 650\ 000/38.7 \times 0.65 = 65\ 593 \text{ m}^3 \text{ of natural gas}.$$

It is apparent that there are a number of issues in the solution which could be challenged. In the absence of information specifically relating to the estimate, one must make value judgements based, of course, on common sense and engineering experience. This is one of the roles of the professional engineer.

The presentation of the fuel estimate to the client or senior management must include the qualifications that have been used in order to justify the estimate. The qualifying statements for this solution are made below.

4.6 Qualifying remarks – sports centre

- Maximum occupancy at the centre at any one time is 330, excluding spectators.
- No provision for hot water has been made for spectators.
- A diversity factor of 40% has been applied to the use of the facilities. If the swimming pool alone is taken as being in full use the annual fuel estimate increases to 75 751 m^3. (Do you agree?) This calculation, whilst not illustrating a practical possibility, may assist in putting the annual estimate into context.
- Some of the figures for the consumption of hot water for occupants using the facilities have been taken from recommended guidelines. In the absence of data other figures have been estimated.
- The average consumption of hot water per person at the centre, based upon the data adopted, is 45 460/2020 = 22.5 litres.
- This annual fuel estimate provides one of the base loads for the centre.
- The fuel estimate is for the provision of hot water. It does not include the fuel required to heat up water resulting from evaporation from the swimming and toddlers' pools. Refer to Chapter 7.

4.7 An alternative method of estimating

Energy consumption arising from hot water supply can be estimated from knowing the floor area of the building. The *CIBSE Building Energy Code* [2] tabulates the mean power requirements of hot water supply in W/m^2 for various buildings and this is reproduced in Table 4.3. Table 4.3 is based upon the provision of hot water supply by the space heating boiler.

Table 4.3 HWS provision in W/m^2 of floor

Building type	HWS (W/m^2)
Office: five- or six-day week	2.0
Shop: six-day week	1.0
Factories:	
five-day single shift	9.0
six-day single shift	11.0
seven-day multiple shift	12.0
Warehouses	1.0
Residential	17.5
Hotels	8.0
Hospitals	29.0
Education	2.0

Annual energy consumption for HWS derived from Table 4.3 is:

$q \times A \times S \times 24 \times 3600/106$ MJ.

where A is the floor area in m^2 and S is the number of working days under review.

The units of the terms in the formula, where W/m^2 = J/s m^2, are:

$(J/s\ m^2) \times m^2 \times days \times hr/day \times sec/hr = J.$

Thus, converting the solution to MJ and reducing the numbers in the formula to a single constant:

$$AED = 0.0864q \times A \times S \text{ MJ.} \qquad (4.5)$$

There now follows an example adopting equation (4.5).

Example 4.4
A hotel has a floor area of 1100 m^2. Estimate the annual energy demand for hot water supply given that it is open for eleven months of the year.

Solution
From Table 4.3 demand for HWS for a hotel is given as 8 W/m^2. If the hotel is in use for 48 weeks in the year, then by adopting equation (4.5) for the annual energy demand:

$AED = 0.0864 \times 8 \times 1100 \times 7 \times 48 = 255$ GJ.

This assumes that the hotel is filled to capacity at all times, which is unlikely. A utilization factor should be applied here to arrive at the final estimate. You should now list the factors which will affect the utilization factor. A utilization factor of 0.5 assumes that the hotel is 50% full throughout the eleven months and will be applied here. Thus:

$AED = 128$ GJ.

The annual energy demand per square metre for HWS, assuming a 50% utilization factor, will be: $128/1100 = 0.116$ GJ/m^2.

You will see from Chapter 6 that the Standard Performance Indicator for hotels is 1.3 GJ/m^2 for a **good** classification, assuming no corrections are applicable. It is noteworthy that the energy for HWS in this example is about 9% of this figure.

Have a look at the hot water requirements in case study 3.4.

The annual energy consumption for the provision of a hot water supply will vary in proportion to the total consumption depending upon the use to which the building is put. It would be helpful to have a comparison with the alternative method of estimating the annual energy demand for HWS. However, to do this it is necessary to know the number of guests that the hotel can house. Adopting an arbitrary figure of 20 m^2 of floor per person, which allows for reception, dining and lounge areas, as well as the bedroom and access ways in the hotel, the number of guests at the hotel when it is full will be $1100/20 = 55$. Adopting the earlier formula, equation (4.1):

$$AED = 231 \; YN(1 + z/100)EF \quad kJ$$

and taking consumption per person from Table 4.1 as 40 litres:

$$AED = 231 \times 55 \times 40(1+10/100) \times 7 \times 48 = 188 \text{ GJ}.$$

This compares with 255 GJ which is determined from the floor area and Table 4.3. The discrepancy is apparent here and reinforces the need for caution if historical data is not available.

4.8 Chapter closure

You now have the skills to estimate the annual energy/energy costs for the provision of hot water in four categories of building. The extension of these acquired skills to other building types is a matter of applying the principles learnt in this chapter. It requires reference to current data, access to historical data specific to the building type, if available, and the application of good engineering common sense. Remember the importance of qualifying the annual fuel estimate.

Energy consumption for cooling loads 5

a	absorption coefficient
AEC	annual energy consumption (kWh)
AED	annual energy demand (kWh)
COP	coefficient of performance
dt	design temperature difference (K)
EH	Equivalent Hours of operation at full load
I_t	intensity of direct + diffuse solar radiation on outer surface (W/m^2)
MSDD	maximum SDD
R_{so}	outside surface resistance (m^2 K/W)
S	number of days under review
SDD	Standard Degree Day(s)
t_{ai}	indoor air temperature
t_{ao}	outdoor air temperature
t_b, B	base temperature in °C
t_{ei}	indoor environmental temperature
t_{eo}	sol-air temperature
t_m	mean daily outdoor temperature
t_n	minimum outdoor air temperature
t_x	maximum outdoor air temperature

5.1 Introduction

It is not the purpose of this chapter to show how cooling loads are determined. Recourse can be made to another publication in the series for detailed design procedures. There are four main factors to be accounted for in the design of the maximum cooling load for an occupied building. They are:

- solar heat gain through glazing which is instantaneous;
- conductive heat gain through the opaque envelope of the building; this will be cumulative over time in a building having a medium or high thermal inertia;
- infiltration of outdoor air;
- internal heat gains such as sensible and latent heat gains from the occupants, and heat gains from equipment and lighting.

To obtain a reasonably reliable estimate of the cumulative effect of conductive heat gain to a given building it is necessary to use computer-based dynamic simulation.

The cooling load for solar and conductive heat gains is calculated from peak conditions in the summer when the refrigeration plant may be required. Low altitude solar heat gains through glazing in the winter months can be offset by introducing fresh air approaching ambient temperature to reduce the use of the refrigeration plant. On this basis the cooling load for solar and conductive heat gains in the UK can therefore be considered seasonal.

The cooling load required to offset internal heat gains, on the other hand, is present at least while the building is occupied and can therefore be considered as the base load. In the winter, however, the temperature rise indoors after occupation of the building resulting from the internal heat gains will offset the heating load. The demand for the base load cooling will therefore be reduced if not cancelled. The total cooling load for a building is therefore the sum of the seasonal load and the base load.

The determination of annual energy costs in the UK related to air conditioning plant providing sensible and latent cooling to buildings is not well documented. The basis upon which estimates of the seasonal cooling load could be made for external heat gains is from the use of cooling Degree Day figures which can be purchased from the Meteorological Office for the same geographical locations as those for heating Degree Days.

There are however limitations to the application of cooling Degree Days when they are associated with solar and conductive heat gains:

- they are based upon maximum and minimum outdoor **air** temperature;
- the air conditioning cycle can include reheating the air after dehumidification and this is not accounted for;
- the extent of free cooling, like, for example, using ambient air ventilation at night to cool the building structure, is not accounted for;
- as with heating Degree Days, the effects of solar radiation are not accounted for although they are acknowledged in the determination of the seasonal design cooling load for the building;
- the effects of shading and hence building shape and orientation are not accounted for although these factors are acknowledged in the determination of the seasonal design cooling load to offset solar heat gains.

The estimate of the annual energy requirements for the cooling base load to offset internal heat gains must also be included. This will usually depend upon the extent of the occupancy period.

COOLING DEGREE DAY DATA

The following data relating to cooling Degree Days was obtained from the Meteorological Office for one locality in the UK [1]: 'The annual 20 year average from 1975 until 1994, for London Heathrow (Thames Valley) for standard cooling Degree Days above 15.5°C is 29. The only months in which this figure is zero are January, February and December.' This implies that cooling Degree Days have occurred during every other month of the year over the 20-year period under review.

The current emphasis on energy conservation encourages plant operators to consider running the refrigeration plant only during the summer months unless there is a specific requirement for maintaining a constant relative humidity and dry bulb temperature. The estimation of annual energy consumption forecasts for the seasonal cooling load are made difficult on the basis of Meteorological Office cooling SDDs.

Figure 5.1 is taken from the *Fuel Efficiency Booklet No. 7* and shows how cooling Degree Days are calculated [2].

The actual number of standard cooling Degree Days for a given location are assessed using maximum t_x and minimum t_n daily outdoor temperatures about a base temperature of 15.5°C. When t_x and t_n are both above base temperature

$$SDD/day = 0.5(t_x + t_n) - 15.5.$$

When t_x is above base temperature by more than t_n is below

$$SDD/day = 0.5(t_x - 15.5) - 0.25(15.5 - t_n).$$

When t_n is below base temperature by more than t_x is above

$$SDD/day = 0.25(t_x - 15.5).$$

The maximum possible number of SDDs annually is determined from:

$$MSDD = S(t_x - 15.5),$$

where S is the number of days cooling, taken as 275 assuming that no cooling Degree Days are recorded in the UK for the months of December, January and February.

The following example illustrates how standard cooling Degree Days are calculated on a daily basis.

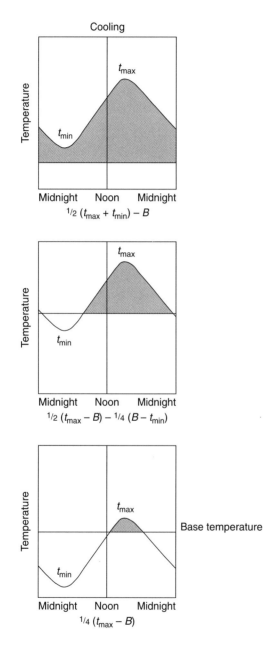

Figure 5.1   Calculation of cooling Degree Days.
Source: *Energy Efficiency Booklet No. 7.*

Example 5.1
Determine the number of SDDs in the sample week taken from data recorded in the Thames Valley during the summer of 1995. What is the maximum number of SDDs for the period under review?

Data:

Day	t_x	t_n
Monday	30	18
Tuesday	28	17
Wednesday	25	16
Thursday	20	13
Friday	18	10

Solution

Day	Calculation	SDD/day
Monday	0.5(30 +18) − 15.5	8.5
Tuesday	0.5(28 +17) − 15.5	7.0
Wednesday	0.5(25 +16) − 15.5	5.0
Thursday	0.5(20 − 15.5) − 0.25(15.5 − 13)	1.6
Friday	0.25(18 − 15.5)	0.6

Actual number of SDD from Monday to Friday = 22.7
MSDD = 5(30 − 15.5) = 72.5

COMMENTS ON SOLUTION TO EXAMPLE 5.1

- The cooling SDDs are high due to the exceptional summer of 1995.
- MSDD is calculated so that a comparison with the recorded SDD can be made. Note that t_x was taken as the maximum daily outdoor temperature in the period under review.
- An approximate method for calculating SDDs for each day may be obtained from:

    SDD/day = $0.5(t_x + t_n) − 15.5$.

By this method the SDD for the period under review comes to 20.0. Do you agree? An alternative method for finding the mean daily outdoor temperature in the process of determining cooling SDDs might be by calculating the average monthly mean daily temperatures from the monthly heating Degree Days which are published by the Meteorological Office and are freely available [3].

The formula is used in the notes to Example (1.5), and is adapted for use here:

Table 5.1 Calculation of average monthly mean daily temperature

Month	Heating SDD	Calculation	t_m
May	114	$114 = 31(15.5 - t_m)$	11.82
June	50	$50 = 30(15.5 - t_m)$	13.83
July	21	$21 = 31(15.5 - t_m)$	14.82
August	24	$24 = 31(15.5 - t_m)$	14.73
September	55	$55 = 30(15.5 - t_m)$	13.67

$$\text{SDD/month} = S(15.5 - t_m)$$

where S is the number of days in the month under review.

Taking the published average heating SDD in the Thames Valley for May, June, July, August and September for the 20 years to 1995, the average monthly mean daily outdoor temperatures t_m are shown in Table 5.1.

It is quite apparent from Table 5.1 that all the monthly mean daily outdoor temperatures are below the standard base temperature of 15.5 °C and therefore will not yield any cooling Degree Days. Have a look again at Example 5.1, in which t_m is greater than t_b on each day of the period under review.

The reason for the apparent requirement for no cooling during the summer months highlighted in Table 5.1 is because by taking monthly averages the calculation of t_m is not sufficiently sensitive. For the calculation of cooling SDDs it is therefore necessary to use **daily** mean daily outdoor temperatures to determine cooling SDD/day.

There is in fact an argument for determining the daily t_m over 12 hours of daylight rather than 24 hours for the cooling SDD because of the distortion caused by the fall in ambient temperature which can occur overnight.

ANNUAL ENERGY DEMAND FOR SEASONAL COOLING

The annual energy demand for a given seasonal design cooling load to offset the effects of external heat gain can be calculated from:

$$\text{AED} = \text{design seasonal cooling load in kW} \times 24\text{SDD}/dt \text{ kWh} \quad (5.1)$$

where dt is the design outdoor temperature t_{ao} minus the design indoor temperature t_{ai}:

$$dt = t_{ao} - t_{ai}$$

and $\text{EH} = 24\text{SDD}/dt$ are the Equivalent Hours at full load. See Chapter 2.

The annual energy consumption for electrically operated vapour compression refrigeration central cooling plant is:

AEC = 0.5(AED) kWh.   (5.2)

The annual energy consumption for similar distributed cooling plant is:

AEC = 0.78(AED) kWh.   (5.3)

The constants for converting AED to AEC are taken from the *BSRIA Rules of Thumb* [6] and account for the coefficient of performance (COP) of the refrigeration plant.

The example which follows is based upon the cooling SDD issued by the Meteorological Office for a building subject to cooling.

Example 5.2
Determine the annual energy consumption for a centralized design seasonal cooling load of 100 kW which offsets the external heat gains to a building located in the Thames Valley given that the design dry bulb temperatures are 30°C outdoors and 23°C indoors.

Solution
From equation (5.1):

AED = 100 × 24 × 29/(30 − 23) = 9943 kWh.

From equation (5.2):

AEC = 0.5 × 9943 = 4972 kWh.

QUALIFYING REMARKS ON EXAMPLE 5.2

- The question and solution do not account for internal heat gains in the building.
- The solution accounts for the effect of high outdoor dry bulb temperature but does not account for the effect of sol-air temperature.
- It accounts for 29 SDDs which in any year could be spread over nine months according to the qualifications relating to this figure issued by the Meteorological Office. Unless the 29 SDDs occurred over consecutive periods it is unlikely that the cooling plant would be triggered to offset external heat gains.
- If the cooling SDDs do occur over consecutive periods it is assumed that the plant will be triggered when the indoor temperature has risen to the set point.
- It does not account for free cooling.

5.3 Factors affecting the estimation of AED for seasonal cooling

As indicated in the Introduction, there are a number of factors which will affect the annual estimate of energy required to offset solar and conductive heat gains in the summer.

SOL-AIR TEMPERATURE

Sol-air temperature t_{eo} is defined as the outdoor air temperature which in the absence of solar radiation would give the same temperature distribution and rate of heat transfer through an external wall or roof as exists with the actual outdoor air temperature and the incident solar radiation. It therefore accounts for the high temperatures achieved on the external surfaces of buildings exposed to solar radiation. By definition it also accounts for building orientation since listings are given for vertical and horizontal surfaces.

The effects inside the building on mean radiant temperature are delayed and depend upon the construction of the building envelope. For buildings with massive walls, for example 0.5 m or more of stone, with the outer surface painted a brilliant white and with small areas of recessed glazing, the thermal inertia is considerable and the thermal response factor is high, and the effects of a day of high intensity solar radiation will not be felt indoors until the evening.

Conversely a steel frame building with concrete floors and light-weight external wall panels will have a very low thermal inertia and hence thermal response, which will ensure that the effect on mean radiant temperature indoors from solar radiation incident upon the external walls will be almost immediate.

In its refurbishment programme for domestic buildings Germany has used external cladding which has the effect of delaying the increase in mean radiant temperature indoors in the summer until the early evening.

The sol-air temperature can be calculated from:

$$t_{eo} = t_{ao} + (R_{so} \times a \times I_t) \ °C,$$

where

R_{so} = 0.05 m^2 K/W for walls
    = 0.04 m^2 K/W for roofs
a = 0.3 for light coloured surfaces, 0.6 for medium coloured surfaces and 0.9 for dark surfaces.

The *CIBSE Guide* gives values of sol-air temperatures for horizontal and vertical surfaces at latitude 51.7 degrees north for each month of the year assuming a clear sky [4]. If one takes the mean sol-air temperatures for vertical surfaces from the data in the *CIBSE Guide* it is possible to determine the cooling SDD based upon these values. Table 5.2 shows the results adopting the appropriate formula from

Table 5.2 Annual cooling SDD based on mean sol-air temperature

Month	Mean sol-air max	min	Calculation	Cooling SDD
May	21.5	14.9	31(18.2 − 15.5)	84
June	24.8	18.8	30(21.8 − 15.5)	189
July	27.4	20.8	31(24.1 − 15.5)	267
August	25.4	18.2	31(21.8 − 15.5)	195
September	21.4	14.3	30(17.9 − 15.5)	72
			Annual total	807

Section 5.2. and substituting sol-air temperature for mean daily outdoor temperature.

If the solution to Example 5.2 is now based upon the annual total of cooling SDD in Table 5.2 the result is as follows.

5.4 The use of sol-air temperature for estimating AED for seasonal cooling

ALTERNATIVE SOLUTION TO EXAMPLE 5.2

From equation (5.1):

AED = design seasonal cooling load in kW × 24SDD/dt.

When using the sol-air temperature, t_{eo}, design d$t = (t_{eo} − t_{ei})$.

The environmental temperature indoors, t_{ei}, will be marginally higher than t_{ai}, due to the relatively high value of the mean radiant temperature indoors resulting from solar heat gains. If this is discounted:

AED = 100 × 24 × 807/(30 − 23)
    = 276 686 kWh.

From equation 5.2:

AEC = 0.5 × 276 686 = 138 343 kWh.

QUALIFYING REMARKS ON ALTERNATIVE SOLUTION TO EXAMPLE 5.2

- The question and solution do not account for internal heat gains in the building.
- There is clearly a significant difference between the two solutions to Example 5.2. The first solution is not realistic at all. The second is based upon fixed values of the sol-air temperature for latitude 51.7 degrees north for five months of the summer. It does not account for variations in cloud density and ambient air tempera-

ture and the consequent effect on sol-air temperature from year to year.

- It is assumed that plant operation is triggered when indoor temperature reaches the set point.
- It does not account for free cooling.
- The potential requirement for sensible heating to the supply air after dehumidification is not accounted for in the assessment of annual energy consumption.
- The use of sol-air temperature to estimate the annual energy consumption for solar heat gains needs to be tested and validated from historical data. It is clear, for example, that the annual SDD of 807 is a fixed value at latitude 51.7 degrees north and does not vary with annual changes in climate as the cooling SDD based upon mean daily outdoor air temperatures does.

CONDUCTIVE HEAT GAINS THROUGH THE OPAQUE BUILDING ENVELOPE

Like heat input to a building during the winter prior to occupation, conductive heat gains resulting from solar radiation take time to penetrate the building envelope. The length of time is dependent upon the thermal inertia of the building and hence its thermal response. It is also dependent upon the location of the thermal insulation sandwich within the building envelope. It was argued that for winter heating the ideal location for an intermittently occupied building was at the inside surface. See Chapter 2.

For conductive heat gains in the summer it can be argued that the best location for the thermal insulation is at the outside surface. This will reduce the impact of solar heat gains indoors and will have a beneficial effect on the energy consumption required for seasonal cooling. It should come as no surprise, therefore, to frequently find the insulation sandwich located midway in the external walls of buildings in temperate climates.

Conductive heat gains through the opaque element of the building envelope are accounted for in the determination of the maximum design cooling load to offset solar heat gains. Since they are usually cumulative over time this is best determined from reliable computer based software.

SOLAR HEAT GAIN THROUGH GLAZING

Heat gain resulting from solar radiation through glazed windows is instantaneous. The *CIBSE Guide* gives cooling loads in W/m^2 of glass due to solar gain through vertical glazing at a variety of latitudes and for eight compass orientations for each month of the year [5].

The *BSRIA Rules of Thumb* gives approximate figures of 150 W/m² of glass for south facing windows from June to September and 250 W/m² for east/west facing windows assuming in both cases that internal blinds are in use [6]. This is accounted for in the determination of the maximum cooling load to offset solar heat gains. The peak seasonal heat gain occurs when the sum of the instantaneous heat gain through the glazing and the heat gain through the opaque portion of the building envelope is a maximum.

SHADING

Shading from the effects of direct solar radiation can take a number of forms:

- shade provided by the effect of recesses in the external envelope of the building;
- shade provided by static or movable external blinds;
- transient shading provided by the orientation of the building on one or more of its external walls;
- permanent or transient shading provided by surrounding buildings, screens or vegetation.

Permanent fixed and movable shading should be accounted for in the calculation of the seasonal cooling load to offset solar heat gains.

REDUCING EXTERNAL HEAT GAINS

For a building whose external envelope is either massive or well insulated thermally, with windows recessed and/or shaded from direct solar radiation, the peak indoor mean radiant temperature will be delayed until the evening. If the building is unoccupied at night, like an office block for example, the seasonal cooling load can be reduced by the provision of full fresh air mechanical ventilation at night to cool the building structure. The external envelope, of course, must be designed to delay the effect of the outdoor heat gains on the mean radiant temperature indoors until the evening. If this is done the refrigeration plant will normally only be required to offset the instantaneous solar heat gains through the glazed windows. With suitable external blinds and choice of glass the peak seasonal cooling load for the building can be reduced substantially.

You can see that in this scenario the building structure requires designing with energy conservation in mind. This is not the case with many buildings constructed before 1985 and so the energy manager may well be responsible for an air conditioned building having a substantial cooling seasonal and base load, a significant proportion of

Table 5.3 Approximate data for the calculation of indoor heat gains

Internal heat gain	Load/unit floor area (W/m^2)
Sensible and latent heat gains from the occupants (metabolic)	20
Lighting	10–25
Office equipment	20–40
Small power	5

Source: *BSRIA Rules of Thumb* [6].

which is required to offset the effect of the seasonal load which is attributable to solar and conductive heat gains.

The effect of the accumulation of conductive heat gains over time through the opaque envelope of a building will mean that the mean temperature of the structure at the beginning of each day may slowly rise for an intermittently occupied building.

5.5 Estimation of AED for the cooling base load

The cooling base load is calculated from the heat gains within the building. The BSRIA Rules of Thumb gives an approximate guide, based on unit floor area [6]. Table 5.3 lists some of this data.

The heat gains from occupants is better estimated from the number normally in the building and upon the activity, since it can vary from a total of 100 to 400 W per person.

The *CIBSE Guide* gives more detailed data [7] which should be used in preference to the *Rules of Thumb*.

There has been a substantial development in energy efficient fluorescent lighting recently using slimline tubes of 16 mm diameter. Claims of 25% less energy consumption at 500 lux illuminance [8] are made and this will clearly have a significant effect upon heat gain per unit floor area attributed to lighting and hence the cooling base load.

The energy manager should consider checking the illuminance levels in the buildings for which he or she is responsible to ensure that they are not excessive for the activities and tasks being undertaken. Appendix 7 tabulates standard service illuminance for various activities and tasks. Current office equipment such as computers, printers, etc. tend to consume less power and hence contribute less to indoor heat gains than models even five years old.

ANNUAL ENERGY DEMAND FOR BASE LOAD COOLING

The formula for AED for base load cooling is:

AED = design cooling base load in kW × annual hours of
   operation in kWh.   (5.4)

The annual energy consumption, AEC, may be calculated from either equation (5.2) or (5.3).

The case study which follows illustrates the calculation of AED for heat gains adopting a mean sol-air temperature and the effects of internal heat gains on the winter heating demand.

5.6 AED – an untested alternative estimate

Case study 5.1

A five-storey office block has a treated floor area of 450 m^2 per floor and has an occupancy of 200. The heat gains from the lighting are 12 W/m^2 and those from the office desk top computers which are evenly spread throughout the building are 20 W/m^2. The cooling plant for both the seasonal and base loads is situated locally with air cooled condensers positioned on the roof.

If the heat gains from the occupants are 100 W per person and occupation of the offices is from 09.00 to 18.00 hours, five days a week, 50 weeks a year, determine the cooling base load for the building.

Given that the seasonal cooling load to offset peak solar and conductive gains to the building is 80 kW and summer design indoor and outdoor air temperatures are 23°C and 28°C, respectively, estimate the annual energy consumption for total cooling in the building using SDDs derived from the sol-air temperature for the seasonal cooling load. If the design sensible heat loss for the building is 120 kW show the cooling and heating profiles graphically.

SOLUTION

The sensible and latent heat gains from the occupants are
    $200 \times 100 = 20$ kW.
For permanent artificial lighting, the heat gain is
    $12 \times 450 \times 5 = 27$ kW.
Heat gains from office equipment are
    $20 \times 450 \times 5 = 45$ kW.
The total base load is therefore 92 kW.
From equation (5.4):

   AED $= 92 \times 9 \times 5 \times 50 = 207\,000$ kWh.

From equation (5.3):

   AEC $= 0.78 \times 207\,000 = 161\,460$ kWh.

From equation (5.1) the AED for the seasonal cooling requirement will be:

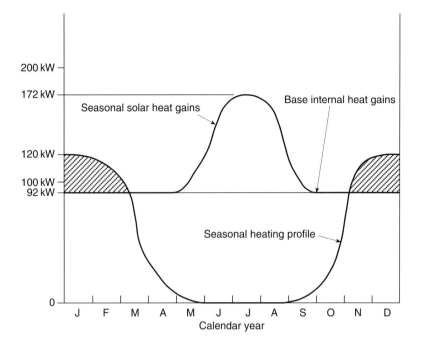

Figure 5.2 Profiles of heat gains after occupation for case study 5.1.

AED = seasonal cooling load in kW × 24SDD/d*t*
     = 80 × 24 × 807/(28 − 23)
     = 309 888 kWh.

From equation (5.3):

AEC = 0.78 × 309 888 = 241 713 kWh.

The estimated annual energy consumption for both seasonal and base load cooling is:

AEC = 161 460 + 241 713 = 403 173 kWh.

Figure 5.2 shows the seasonal solar heat gains profile, the internal heat gains profile and the heating profile over a calendar year. Note that the indoor heat gains will occur only after occupation of the premises.

Figure 5.3 shows the cooling demand profile to offset solar heat gains and heat gains indoors after occupation of the building. It can be seen from the graph that the internal heat gains will offset some of the heating demand and that the design maximum heating demand which results will be 120 − 92 = 28 kW after occupation of the premises.

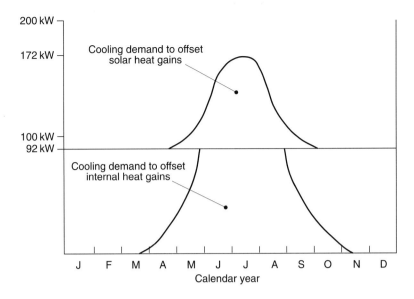

Figure 5.3 Cooling demand profile for case study 5.1.

QUALIFYING REMARKS FOR SOLUTION TO CASE STUDY 5.1

- In the winter the design heat load of 120 kW will be required to bring the building up to temperature before occupation.
- The full cooling base load of 92 kW will in the event only be required during the summer months. This will affect the AED estimate.
- Indoor heat gains in the winter will offset some of the heating load.
- An article in the *Building Services and Environmental Engineer* [9] relating to indoor heat gains claimed that internal heat gains for office equipment in buildings erected in the 1980s at 30 W/m² was excessive and that the actual gain was nearer 7 W/m². The article also claimed that the population density which was taken as 7.5 m²/person was found in the event to be as low as 27 m²/person. Clearly, up-to-date information relating to these heat gains may be required.
- The use of the mean sol-air temperature in the determination of Degree Days produces a fixed annual value which is related to the position of latitude. It does not account for the effects of cloud cover variations annually. It has not been tested or validated from historical data.
- The effects on variation in base load cooling to offset indoor heat gains during the winter season could be predicted from tables of weekly/monthly heating Degree Days.

- It is left to you to list further qualifying remarks relating to estimating AEC for vapour compression refrigeration cooling plant in occupied buildings. What information would be needed to generate a thermal performance line for refrigeration plant responding to solar heat gains and heat gains indoors?

The dependent variable will be energy consumption and the independent variable will be SDD or outdoor mean daily temperature. Have a look at Chapter 10.

5.7 Chapter closure   You have been introduced to cooling Degree Days issued for one locality by the Meteorological Office and how they are calculated. You know the limitations of applying this data to the estimation of cooling loads for occupied buildings. An introduction to an alternative untested model using the mean sol-air temperature for the estimation of seasonal cooling loads offers you a comparison which is illustrated in case study 5.1.

Performance indicators 6

AEC   annual energy consumption
AED   annual energy demand
CDY   carbon dioxide emission yardstick
ECY   energy consumption yardstick
EEO   Energy Efficiency Office
PI   performance indicator
SCT   standard consumption target
SPI   standard performance indicator

Nomenclature

6.1 Introduction

Local authorities and other owners of large numbers of widely dispersed buildings and even owners of buildings on single sites have the opportunity of employing yardsticks for the purpose of comparing the performance of buildings with similar characteristics. These provide one of the checks and balances which help to promote authenticity and validation in the preparation of accounts for building stock to auditors and shareholders.

If, for example, the maintenance costs on a particular building type prove to be excessive when compared with another building being used for the same purposes and located in a similar environment, or when compared with appropriate historical data, it provides evidence for future policy action. If no comparison is made no one is any the wiser and any additional costs continue to accumulate.

Energy consumption in buildings is considered in the same manner and buildings having similar usage and occupancy patterns can be compared. Historical data is now available in the form of tables of performance indicators which express annual energy consumption per square metre of floor at different levels of acceptance and for different building types. Energy consumption yardsticks and carbon dioxide yardsticks are published by the Energy Efficiency Office in their *Introduction to Energy Efficiency* booklets for various building usages. Refer to Appendix 4. Standard performance indicators are published by CIBSE [1]. The latter will be adopted in this chapter with a comparison made later on.

6.2 Performance indicators

The performance indicator for a building is calculated from:

PI = AEC/treated floor area.

Treated floor area includes all areas excluding plant rooms and other areas not heated.

The annual energy consumption (AEC) refers to the historical data from fuel invoices.

For a projected building fuel invoices are not available and the annual energy demand (AED) can be determined from design data (see Chapters 2, 3 and 4) and the estimated annual energy consumption is

AEC = AED/seasonal efficiency.

The units for measurement are usually GJ/m^2 or kWh/m^2.

Note that 1 kWh = 0.0036 GJ; therefore 1 GJ = 278 kWh.

You should now have a look at the case studies in Chapter 3 and calculate the PI for each.

Clearly, calculated performance indicators must be compared with care with the published indicators or yardsticks since they will depend upon many factors. From Chapter 1 it is apparent that performance indicators will be affected by the annual Degree Day totals which vary from one location in the UK to the next and which are the main contributory factor in determining AED for a projected building project.

Likewise for an existing building where historical fuel accounts can be used to arrive at a performance indicator there will be differences arising from geographical location. In addition to geographical location the performance indicator for buildings having similar usage will be affected by:

- building size and shape;
- level of engineering services;
- occupancy pattern;
- level of energy required for manufacturing processes.

The standard performance indicators, SPIs, published by CIBSE therefore need to be corrected before being compared with the performance indicator determined for the building under investigation.

After correction the SPI is termed the standard consumption target, SCT, and it is this which is compared with the performance indicator found for the building under scrutiny.

6.3 Standard performance indicators for some common building types

Historical data relating to energy use in various buildings provides useful benchmarks for comparisons to be made by building owners and energy managers. Performance indicators published by CIBSE [1] are based upon levels of energy use in buildings in the 1970s and are reproduced in Table 6.1.

Table 6.1 Standard performance indicators (SPIs)

Building type	Occupancy	PI for stated classification (GJ/m²)				
		Good	Satisfactory	Fair	Poor	Very poor
Office	single shift, 5 day week	<0.7	0.7–0.8	0.8–1.0	1.0–1.2	>1.2
Factories	single shift, 5/6 day week	<0.8	0.8–1.0	1.0–1.2	1.2–1.5	>1.5
Warehouses	single shift, 5/6 day week	<0.7	0.7–0.8	0.8–0.9	0.9–1.2	>1.2
Schools	single shift, 5 day week	<0.7	0.7–0.8	0.8–1.0	1.0–1.2	>1.2
Shops	single shift, 6 day week	<0.7	0.7–0.8	0.8–1.0	1.0–1.2	>1.2
Hotels	continuous, 7 day week	<1.3	1.3–1.5	1.5–1.8	1.8–2.2	>2.2

The data set out in Table 6.1 applies to heating and naturally ventilated buildings with artificial lighting, normal use of electricity and having normal hours of occupancy. Electricity for power drives and machines as would be found in manufacturing is not included.

Energy consumption yardsticks given in the EEO's *Introduction to Energy Efficiency* Booklets (refer to Appendix 4) are split into fossil fuel consumption and electricity consumption.

Part four of the *CIBSE Building Energy Code* gives a further table of more detailed SPIs for offices. This table also differentiates between the site consumption of fossil fuel and electricity.

CLASSIFICATION OF LEVELS OF ACCEPTANCE

The classification of performance is separated into five levels of acceptance, namely: good, satisfactory, fair, poor and very poor. The CIBSE [1] identifies the classifications as follows:

Good low energy consumption; indicative of careful control and good energy management procedures;

Satisfactory energy usage consistent with sensible operating procedures;

Fair barely average performance; for typical situations significant savings should be achievable;

Poor energy usage is high; for typical situations indicative of significant heat losses in winter and/or poor control of energy use;

Very poor energy usage is excessive; immediate action should be taken to investigate and remedy.

Correction factors have been adapted from data published by the CIBSE [1] and are given below in Table 6.2. These must be applied with great care.

There are four qualifications, three of which apply to Table 6.2. These qualifications are now identified.

6.4 Correction factors for standard performance indicators

Table 6.2 Correction factors for standard performance indicators

Item	Specification	Correction factor
Occupancy period	Single shift, 7-day week	1.2
	Double shift, 5-day week	1.3
	Double shift, 7-day week	1.4
	Continuous	1.4
Engineering services	Air conditioning	1.4
	Mechanical ventilation	1.3
	Electrical heating	0.8
	Purchased heat	0.8
	Space heating by gas, oil or coal	1.0
Location	Scotland or very exposed sites	1.1
	South-west England	0.9
	Remainder of United Kingdom	1.0
Size	floor area $< 5000 \text{ m}^2$	1.0
	floor area $> 5000 \text{ m}^2$	0.95
	floor area – very large buildings	0.90

(1) CORRECTION FACTOR FOR ENERGY USED IN PROCESS AND
MANUFACTURING

The energy consumption for process and manufacturing is assumed to
consist of no more than 20% of the energy consumption for building
services in the preparation of the SPIs in Table 6.1. If it exceeds this
proportion an adjustment should be made to the PI for the building
before a comparison is made.

Example 6.1
A factory located near Manchester has a treated floor area of 900 m²
and works a six day week, single shift and has the following annual
energy consumptions:

space heating AEC = 800 GJ
process and manufacturing AEC = 350 GJ.

Determine the performance indicator for the factory and compare
it with the equivalent SPI.

Solution
The proportion of energy associated with process and manufacturing is $(350/800) \times 100 = 43.75\%$. This is clearly above the minimum
of 20%.

For the purposes of comparison with the SPIs the total energy
consumption with a maximum of 20% process energy is given by:

space heating + process = min. 80% + max. 20%
                        = 800   + 350 × 20/80 = 888 GJ
PI for the factory   = AEC/floor area = 888/900 = 0.987 GJ/m^2

The PI for the factory can now be compared with the appropriate SPIs.

Referring to Table 6.1 this gives a **satisfactory** classification.

You will note here that the SPIs can be adopted as the STCs since no correction for occupancy, engineering services, location or size is required for this factory.

Another point to make here is that the PI for the factory makes no judgement upon the performance of energy conversion for the manufacturing process.

(2) CORRECTION FACTOR FOR AIR CONDITIONING

The factors for air conditioning and mechanical ventilation having some form of comfort conditioning in Table 6.2 should only be applied where the floor area served by these plants is at least 60% of the total floor area. These factors should be reduced pro rata when the floor area served is less than 60%.

Example 6.2
Determine the correction factor which must be applied to the SPI for a building in which the floor area served by the air conditioning plant is 35%.

Solution
Using the factor for air conditioning from Table 6.2 and taking the factor for space heating as 1.0,

correction factor = (35/60) × 1.4 + (25/60) × 1.0 = 1.233.

(3) CORRECTION FACTOR FOR LOCATION

The SPIs in Table 6.1 have been based upon an annual SDD of 2000. You will note that the annual SDDs listed in Table 1.5 vary from 1840 to 2617. If the Standard Degree Days in the heating season differ by more than +10% or −10% of the product of 2000 × location correction factor, the location factor given in Table 6.2 needs adjustment as follows:

Revised correction factor = annual SDD/2000.

Example 6.3
A building is located at Leuchars in East Scotland where the annual SDD is 2496.

Determine the correction factor which should be applied to the SPIs in Table 6.1 before a comparison is made with the PI for the building.

Solution
From Table 6.2 the correction factor for the location of the building is 1.1.

The product of the location factor and 2000 DD is $1.1 \times 2000 = 2200$.

The percentage difference between this product and the annual SDD for the area is:

$$((2496 - 2200)/2200) \times 100 = 13.45\%.$$

This exceeds the 10% limit.

The correction factor which is applicable in this case for the location of the building will be:

$$\text{correction factor} = 2496/2000 = 1.25.$$

This is due to the severity of the local climate and it is this figure which is used to adjust the SPIs in respect of building location.

(4) CORRECTION FACTOR FOR ELECTRIC SPACE HEATING

The correction factor in Table 6.2 of 0.8 is for all electric space heating. If there is partial electric space heating this factor must be adjusted. For example, given that electric space heating accounts for 20% and oil heating 80% the correction for engineering services will be: $0.2 \times 0.8 + 0.8 \times 1.0 = 0.96$.

The following examples show how the correction factors described above can affect the SPIs when converting them into standard consumption targets (SCTs).

Example 6.4
A four storey office block having a treated floor area of 500 m^2 per floor is located in the Plymouth area. The ground floor is air conditioned and the upper floors are heated by a gas fired boiler serving radiators. The occupancy pattern is single shift seven day week. Adopting the SPIs in Table 6.1 determine the SCTs for the building.

Solution
The correction factor for the occupancy pattern, from Table 6.2, is
1.2.
   The correction factor for the proportion of air conditioning,
which is $(500/(500 \times 4)) \times 100 = 25\%$, will be:

   $(25/60) \times 1.4 + (35/60) \times 1.0 = 1.17.$

Factor for location: from Table 1.5 the annual SDD for the Plymouth area is 1840. This is below the SDD figure of 2000 used in
producing the SPIs. Thus from Table 6.2 the correction factor for
location is 0.9.
   The product of 2000×0.9 is 1800.
   The variation is $((1840 - 1800)/1800) \times 100 = 2.22\%$, which is
below the 10% limit, so the correction factor for location of 0.9 in
the table may be used.
   The correction factor for building size in terms of floor area from
Table 6.2 is 1.0.
   The overall correction to be made to the SPIs will be:

   overall correction $= 1.2 \times 1.17 \times 0.9 \times 1.0 = 1.264.$

Adopting the SPIs from Table 6.1 for offices and tabulating the results:

SPI	Good	Satisfactory	Fair	Poor	Very poor
Office	0.7	0.7–0.8	0.8–1.0	1.0–1.2	1.2
Overall correction	1.264	1.264	1.264	1.264	1.266
SCT	0.88	0.88–1.01	1.01–1.26	1.26–1.52	1.52

The performance indicator calculated for the offices can now be
compared with the SCTs.

Example 6.5
A factory operating a double shift five day week with a treated floor
area of 50 m × 20 m and located near Birmingham has the following
energy consumptions annually:

   Oil fired space heating   900 GJ
   Electric heating   400 GJ
   Process and manufacturing   600 GJ

Determine the performance index for the factory and compare with
the standard consumption targets.

Solution

Total annual energy consumption = 900 + 400 + 600 = 1900 GJ.
Proportion attributed to process = (600/1900) × 100 = 32%.

The energy required for process is outside the 20% limit for use with SPIs.

Total energy consumption with a maximum of 20% process will be:

space heating + process   = minimum of 80% + maximum of 20%
1300   + 1300 × 20/80 = 1300   + 325
                            = 1625 GJ.

The PI for the factory will be

AEC/floor area = 1625/(50 × 20) = 1.625 GJ/m^2.

It is now necessary to convert the SPIs for factories given in Table 6.1 into SCTs for this factory.

From Table 6.2 the following corrections should be made to the SPIs:

Correction for occupancy pattern is 1.3.

Space heating by electricity represents (400/1300) × 100 = 31% of the energy consumption for space heating. The correction for all electric space heating of 0.8 needs adjustment thus:

correction for engineering services = 0.31 × 0.8 + 0.69 × 1.0
                                    = 0.94.

The correction for location:

the SDD for Birmingham, from Table 1.5, is 2357 and the correction for location in Table 6.2 is 1.0. The variation will be ((2357 − 2000)/2000) × 100 = 17.85% . This is greater than the 10% limit.

The revised correction is 2357/2000 = 1.18.
The correction for floor area is 1.0, from Table 6.2.
The overall correction to apply to the SPIs for factories will be:

overall correction = 1.3 × 0.94 × 1.18 × 1.0 = 1.44.

Adopting the SPIs from Table 6.1 for factories and tabulating the results:

SPI	Good	Satisfactory	Fair	Poor	Very poor
Factories	0.8	0.8–1.0	1.0–1.2	1.2–1.5	1.5
Overall correction	1.44	1.44	1.44	1.44	1.44
SCT	1.15	1.15–1.44	1.44–1.73	1.73–2.16	2.16

The PI for the factory was calculated as 1.625 GJ/m^2. This lies within the **fair** classification of SCTs (1.44-1.73) which states: barely average performance; for typical situations significant savings should be achievable.

The Energy Efficiency Office has published a series entitled *Introduction to Energy Efficiency* for a number of building types. See Appendix 4. These publications are at present free of charge and include useful information relating to energy use and energy monitoring. Each booklet includes energy consumption yardsticks for fossil fuels and electricity for low, medium and high rates of consumption expressed in kWh/m^2, as well as yardsticks for low, medium and high emissions of carbon dioxide expressed in kg/m^2.

The performance index calculated for the building, before applying corrections, can be compared with the energy consumption yardsticks quoted in each booklet. The Energy Efficiency Office's energy consumption yardsticks are therefore equivalent to the standard performance indicators used in the CIBSE Code, part 4.

6.5 Further source data for performance indicators

The ECYs separate electricity consumption from site use of fossil fuels whereas it can be assumed that SPIs include both. Typical comparisons between ECYs and SPIs before any corrections are applied are tabulated here. The Energy Efficiency Office's ECYs for a secondary school without a swimming pool, converted to GJ/m^2, along with other building types, are shown in Table 6.3 with the corresponding figures from the *CIBSE Building Energy Code*, Part 4.

The data in each of the Energy Efficiency Office's booklets is more detailed. For example offices are divided into four types and Table 6.4 shows the breakdown.

The monthly *Journal of the Chartered Institution of Building Services Engineering* has recently included in its building analysis articles on new and refurbished buildings, annual energy targets in kWh/m^2 of treated floor [2]. This provides a useful update which accounts for recent changes in the standards of thermal insulation of the building envelope. Table 6.5 gives details of the energy targets from some of the buildings analysed in recent journals.

The European Commission's grant scheme EC2000 funds eight non-domestic buildings, two of which have been completed in the UK.

6.6 Comparison of source data

Table 6.3 ECYs and PIs in GJ/m^2

Building type	Source	Good	Very poor
Secondary school without swimming pool	EEO	<0.62	>0.85
	CIBSE	<0.7	>1.2
Offices	EEO	<0.56	>1.03
	CIBSE	<0.7	>1.2
Factories	EEO	<0.8	>1.17
	CIBSE	<0.8	>1.5
Hotels	EEO	<1.22	>1.94
	CIBSE	<1.3	>2.2

Table 6.4 ECYs for the office

Office type	Low (kWh/m²)	High (kWh/m²)
Smaller office		
fossil fuels	95	200
electricity	36	48
Naturally ventilated open plan		
fossil fuels	95	200
electricity	61	85
Air conditioned open plan		
fossil fuels	100	222
electricity	132	202
Headquarters including computer room and catering		
fossil fuels	132	273
electricity	261	361

Source: EEO

The scheme requires a 50% reduction in energy and carbon dioxide emission compared with traditional buildings and no air conditioning. The buildings in Table 6.5 are East Anglia University Learning Centre and No. 1 Leeds City Office Park.

The Building Research Energy Conservation Support Unit (BRECSU) [3] has reviewed the performance specifications for a new office building and seminar facilities to be constructed at the Building Research Establishment's site at Garston in which performance targets are identified. These are shown in Table 6.6.

In both Tables 6.5 and 6.6 the low annual energy targets (below 100

Table 6.5 Energy targets (kWh/m²) for buildings analysed in the monthly *CIBSE Journal*

Building	Heating	Ventilating	Hot water	Small power	Lighting	Total
Charities Aid Foundation, Kings Hill, Kent	100	30	18	25	45	218
New Scottish Office, Edinburgh	107	refrigeration 9	3	14[a]	26	159
RSPB office, Bedfordshire	–	–	–	–	–	140
Marston Book Services, Oxfordshire	12[b]	ventilation 24	[b]	1.2	1.72	39
Learning Centre, East Anglia University	–	–	–	–	–	95
Elizabeth Fry Building, East Anglia University	26[b]	[b]	[b]	8	16	50
No. 1 Leeds City Office Park	45[b]	16	[b]	48[a]	11	120
Portland Building, University of Portsmouth	[b]		[b]			165

[a] identifies small power, including lifts
[b] identifies services whose energy targets have been combined in the building analysis.

Table 6.6 Performance targets for a new office and seminar facility, BRE

Building type	Annual energy consumption (kWh/m^2)	Annual carbon dioxide emission (kg/m^2)
Narrow plan	Gas 47, electricity 36	34
Narrow plan	All electric 68	46
Deep plan	Gas 47, electricity 43	39
Deep plan	All electric 75	51

kWh/m^2) generally indicate buildings which have been specifically designed as low energy buildings taking advantage of solar heat gains and using natural ventilation for air replacement and night-time cooling.

6.7 Conclusions relating to SCTs and SPIs

The determination of standard consumption targets relies on corrections being made to the standard performance indicators. Both the SPIs and the correction factors result from studies of the levels of energy consumption in different buildings. The studies reveal, as one might expect, considerable scatter. The CIBSE [1] give the reasons for this scatter as variations in:

- maintenance and operational procedures;
- energy management procedures;
- the thermal quality of building and plant.

As with much data that is based on historical evidence it is important to point out that SPIs and hence SCTs should be considered as guides to thermal performance.

If a building achieves a **good** classification this does not therefore imply that a detailed energy consumption survey is not necessary. Similarly a **very poor** classification should not imply that building and plant need complete refurbishment; there may be extenuating circumstances which will only surface following a survey.

6.8 Carbon dioxide emissions

The performance index for a building provides a measure of its energy consumption which can be compared with standard values. The Energy Efficiency Office in its most recent booklets on energy efficiency in various types of building (see Appendix 4) has for the first time included carbon dioxide yardsticks. This is no doubt in response to the international call to reduce levels of carbon dioxide emission. The conversion factors from energy in kWh to kg of carbon dioxide emission annually are based upon 1993 emission factors.

Clearly there is a direct correlation between energy consumption and carbon dioxide emission; a saving in the annual use of energy will

Table 6.7 Conversion of annual energy consumption to emission of carbon dioxide annually

Fuel	Carbon dioxide conversion factor (kg CO_2/kWh energy)
Gas	0.20
Oil	0.29
Coal	0.32
Electricity	0.70

produce a corresponding reduction in carbon dioxide emissions. The conversion factors are reproduced from the Energy Efficiency Office's *Introduction to Energy Efficiency* booklets in Table 6.7.

The emission factor for electricity is based upon fossil fuel power only generation and therefore is not relevant to combined heat and power generating plant or electricity generated by hydroelectric plant.

6.9 Carbon dioxide yardsticks

The Energy Efficiency Office's energy efficiency booklets referred to earlier include yardsticks for carbon dioxide emissions, some samples of which are reproduced in Table 6.8.

Table 6.8 Carbon dioxide emission yardsticks

Building type	Carbon dioxide performance (kg CO_2/m^2)	
	Low	High
School (no pool)	< 46	> 63
Offices	< 62	>100
Factories	< 80	>115
Hotels	<110	>180

There now follows an example in which carbon dioxide emission is estimated and compared with the CDYs.

Example 6.6
A general office in the Thames Valley is used on a single shift, five days per week. It is serviced with a space heating system and a tempered air mechanical ventilation system. From the data, determine the performance indicator for the building and compare it with the standard consumption target. Determine also the annual carbon dioxide emission from the plant and compare it with the CDY.

   Data: annual consumption of gas 1050 GJ, annual consumption of electricity for lighting and power 97 300 kWh, treated floor area 1200 m².

Solution
Clearly much work of a routine nature discussed in other chapters of this book has already been done to arrive at the data given. With reference to Table 6.2 there is no correction for location, occupation pattern or floor area. However there is a correction for mechanical ventilation of 1.3. This correction must be applied to the standard performance indicators given in Table 6.1 for offices. The relevant data for the solution is given in Table 6.9.

Table 6.9 Relevant data for the solution to Example 6.6

Building type	Level of performance				
	Good	Satisfactory	Fair	Poor	Very poor
Office SPI	<0.7	0.7–0.8	0.8-1.0	1.0–1.2	>1.2
Overall correction	1.3	1.3	1.3	1.3	1.3
Office SCT	<0.91	0.91–1.04	1.04–1.3	1.3–1.56	>1.56

$$\text{Annual gas consumption} = 1050 \text{ GJ.}$$
$$\text{Annual electricity consumption} = 97\ 300 \text{ kWh}$$
$$= 97\ 300 \times 0.0036$$
$$= 350 \text{ GJ.}$$
$$\text{Total energy consumption annually} = 1400 \text{ GJ.}$$

Performance indicator for the building = AEC/treated floor area, i.e.

   $PI = 1400/1200 = 1.167$ GJ/m².

Comparing this PI with the standard consumption targets calculated above for this building shows it as having a **fair** performance. You will recall that this classification is described as: barely average performance; for typical situations savings should be achievable. In fact it is likely that during the preparation of an audit potential energy saving measures will be identified.

Determining now the carbon dioxide emission:

Annual gas consumption in kWh = 1050 × 278
$$= 291\ 900 \text{ kWh;}$$
Annual electricity consumption = 97 300 kWh.

From Table 6.7:

CO_2 emission from burning gas $= 291\ 900 \times 0.2 = 58\ 380$ kg;
CO_2 emission from generating electricity $= 97\ 300 \times 0.7 = 68\ 110$ kg;
Total annual emission $= 126\ 490$ k $= 126.49$ tonnes.
Annual emission $= 126\ 490/1200 = 105.4$ kg/m^2.

This compares with the CDY from Table 6.8 of between 62 and 100 kg/m^2 and therefore indicates high carbon dioxide emission. It is apparent that with a **fair** energy classification, carbon dioxide emission may not fall within acceptable levels.

6.10 Chapter closure   You are now able to identify standard performance indicators from source material for different categories of building and apply appropriate corrections to the SPIs for a specific building to arrive at standard consumption targets. Following the determination of the building's performance indicator and taking care in interpreting the results you can advise a client on the projected performance for a new building or the current performance in the case of an existing building and show how it compares against a standard.

It is important to take into account when determining the SCT any local factors which might influence a comparison with the performance indicator calculated for the building being analysed.

You have also been introduced to the estimation of carbon dioxide emissions and the yardstick tables provided by the Energy Efficiency Office in their energy efficiency booklets. See Appendix 4.

Energy conservation strategies 7

Nomenclature

A	actuator
AEC	annual energy consumption
AED	annual energy demand
C	controller
CEM	contract energy management
CV	calorific value
db	dry bulb temperature in °C
d_g	difference in moisture content (kg/kg dry air)
d_h	difference in enthalpy (kJ/kg dry air)
DS	duct stat
DD	Degree Day(s)
dt	design indoor/outdoor temperature difference (K)
E	exhaust air
HWS	hot water supply
ID	immersion detector
M	mass flow rate (kg/s)
M	meter
n_c	heat conversion efficiency
n_s	seasonal efficiency
n_u	heat utilization efficiency
O	outdoor air condition
OD	outdoor detector
PEM	partnership energy management
RS	room stat
S	supply air condition
SDD	Standard Degree Day(s)
SPI	standard performance indicator
v	specific volume (m³/kg)
v_{fr}	volume flow rate (m³/s)

7.1 Introduction

Since the present national average for the purchase of fossil fuel for heating is 4% of turnover, the incentive on the part of senior management to invest in energy saving measures is low with many organizations. The demands upon time are considerable and when cost cutting is required other areas of investigation which offer potentially greater savings than, say, 10% of 4% of turnover take precedence.

However, pressure is now being applied throughout industry and on the domestic scene to reduce the levels of greenhouse gas. Pressure for the reduction of carbon dioxide and oxides of nitrogen may well encourage building owners into action to reduce energy consumption.

The Audit Commission's long-established guidelines on energy management state that for every £1M of fuel purchased one energy manager should be appointed and 10% of the capital expenditure on fuel reinvested in energy conservation measures. This hardly encourages those organizations which have smaller fuel bills. Many such organizations have no energy management policy and those that have give the job to an employee who has other responsibilities. The individual given the part-time job of energy manager needs all the encouragement that he or she can find from top to bottom of the organization.

The large organizations typically have the management of energy consumption as company policy. It is clearly in the interests of the British Airports Authority, for example, with 186 000 m^2 of floor space to have strategies for energy management.

Some large organizations like local authorities and hospitals buy in the energy management expertise in the form of partnership energy management (PEM) [1] or contract energy management (CEM) in which the contractor takes on responsibility for operating and updating the services and plant and shares the savings in energy consumption with the client. With PEM the contractor offers the same services but allows the client to have day-to-day control of plant operation.

7.2 Energy transfer from point of extraction to point of use

For fossil fuels there are a number of stages from extraction to site use which need consideration. Energy conservation strategies apply to every stage although it is at the site use stage that this book is focused.

Figure 7.1 is a block diagram showing the stages from extraction to site use. The non-productive energy use identifies the energy losses sustained in extraction, transport and refinement.

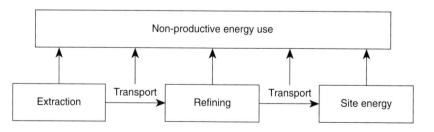

Figure 7.1 Processes in production and distribution of fossil fuel.

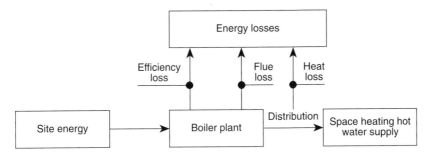

Figure 7.2 Energy losses from point of supply to point of use.

Figure 7.2 is a block diagram showing the energy use on site. The non-productive loss of energy on site is identified as the energy loss due to the inefficiency of the boiler plant and that via the sensible and latent heat losses from the flue gases. If the heating medium transports the heat energy via external ducts there will be losses here also.

Primary energy is that in the fossil fuel at the point of extraction. Site energy is that in the fossil fuel available to the consumer on a site. Table 7.1 which originates from data published by CIBSE [2] lists primary energy fuel factors which relate directly to units of site energy.

Note the size of the primary fuel factor for electricity due to the inefficient conversion of fossil fuels in electrical power only generating plant.

Combined heat and power plant, for example, increases the conversion efficiency from about 30% for power only generating stations to between 50% and 80% for central heating plant. This will have the effect of reducing the fuel factor to between 1.5 and 2.0. Part (f) of the following example shows how the primary energy fuel factor in Table 7.1 is applied.

7.3 Efficiency of space heating plants

There are three types of efficiency related to space heating systems, namely: heat conversion efficiency, utilization efficiency and seasonal efficiency. The *CIBSE Guide* [3] tabulates these efficiencies for different types of plant and system.

Table 7.1 Primary energy fuel factors

Final energy form	Primary energy fuel factor	Remarks
Electricity	3.82	Power only generating plant
Manufactured fuels	1.38	
Oil	1.09	
Natural gas	1.07	
Coal	1.03	

Example 7.1
A factory owner pays £10 000 per annum for light fuel oil for the heating boiler.

(a) How many litres of oil are delivered to site per annum at a cost of 11p/litre?
(b) What is the total energy content of the fuel in GJ/annum?
(c) If the boiler consumes 30 litres of oil per hour, what is its power input in kW?
(d) How many hours does the boiler operate per annum?
(e) If the average efficiency of the boiler between services is 70%, what is the useful power output in kW?
(f) How many units of primary energy does the boiler consume for each unit of useful energy output?

Solution

(a) The quantity of fuel oil burnt = £10 000/0.11 = 9909 litres.
(b) Total energy content of fuel = 90 909 × 40.5 = 3682 GJ.
   Note the CV for light fuel oil is obtained from Table 2.1.
(c) Oil consumption = 30 × 40.5 = 1215 MJ/hour.
   Power = energy/time = 1215 × 1000/3600 = 337.5 kW.
   Power input = 337.5 kW.
(d) Hours of boiler operation = 90 909/30 = 3030 hours/annum.
(e) Useful power output = power input × efficiency = 337.5 × 0.7 = 236 kW.
(f) From Table 7.1 the primary energy content of each unit of site fuel is 1.09.

For each GJ of site energy used 0.7 GJ is useful energy. Thus each useful GJ of energy requires 1/0.7 = 1.43 GJ of site energy. The boiler will therefore consume 1.43 GJ of site energy for each GJ of useful energy. The boiler will therefore consume 1.43 × 1.09 = 1.56 GJ of primary energy from the point of extraction for each GJ of useful energy.

- Heat conversion efficiency, n_c.
  This relates to the boiler plant where n_c = output/input, where output = heat delivered to the system in kW, and input = heat potential in the fuel in kW.
  Heat conversion efficiency should in fact be reasonably steady to a turn down ratio of 30% of full load for conventional boilers. At lower loads boiler efficiency falls away rapidly. It should be noted that heat conversion efficiency is that quoted by the boiler manufacturer and is determined under controlled conditions and not prevailing conditions on site. Heat conversion efficiencies for

different boilers range from 98% for double condensing boilers to 70% for conventional boilers. Refer to Chapter 1, Table 1.2.

- Utilization efficiency n_u.
  This addresses the manner in which heat is emitted where it is needed. It is dependent upon the type of system, the method of control, the disposition and sizing of the space heating appliances, the size, construction and thermal behaviour of the building and the method of operation. It does not account for heat losses from distribution pipework for which separate allowance should be made. n_u = design heat load/heat delivered to the system.

- Seasonal efficiency n_s.
  This is the overall efficiency of boiler plant and system over the heating season and accounts for variations in efficiency over time resulting from the intervals between maintenance and variations in the load on the boiler plant. Typical seasonal efficiencies range from 60% for indirect heating like low temperature hot water to 75% for direct heating like direct fired gas heaters for water or air at the point of use. Refer to Table 1.2.

The relationship between heat conversion efficiency, n_c, utilization efficiency, n_u, and seasonal efficiency, n_s, is expressed as: $n_s = n_c \times n_u$.

7.4 Seasonal and base load demand and consumption

In order to apply energy conservation strategies it is necessary to distinguish between demand and consumption and between the terms seasonal and base load. Seasonal demand and consumption vary with changes in climate over the year and are associated with space heating. Base load demand and consumption relate to a constant and continuous use of energy or fuel as in the case of hot water supply and some manufacturing processes. The difference between demand and consumption can be illustrated by distinguishing between annual energy demand, AED, and annual energy consumption, AEC:

Annual energy consumption = annual energy demand/seasonal efficiency or rearranging:

AED = AEC × n_s.

Figure 7.3 shows a typical plot of seasonal and base load energy consumption for the provision of space heating and hot water supply to a building for twelve months. The plot is made from monthly fuel readings taken on site. In practice there may be small variations in the base load each month. The plot gives a picture of the consumption pattern of fuel over a year. If there is only one metering point for the fuel, the size of the base load can be identified during the summer months when space heating is not required. The base load line can then be extended either side of the summer months on the reasonably safe assumption that it will be more or less constant. The

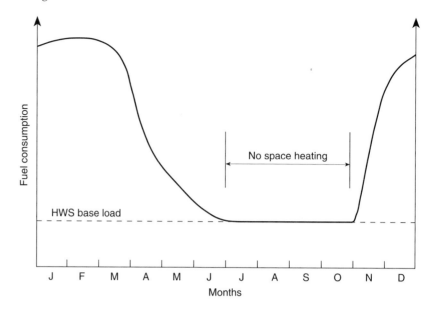

Figure 7.3 Fuel consumption profile for a typical year for space heating and hot water supply.

plot gives little more information and does not show plant performance.

7.5 The energy conservation programme

In order for senior management in an organization to agree a programme of energy conservation measures, it must have good reasons for so doing. In fact there are at least two good motives for action on the part of building owners/occupiers and these will become increasingly effective in time. One is cost saving and the other is to reduce carbon dioxide emissions via the products of combustion into the atmosphere. Legislation will provide the motivation for the latter via grants and penalties. Increases in the tax on fossil fuels will encourage the former.

Manufacturing industry and other large users of fossil fuels in the UK are already some way along the path of energy conservation and the reduction in emissions of carbon dioxide. This is due mainly to the effect of grants and significant achievable savings on fuel costs resulting from deliberate changes in working practices and modernization of energy consuming plant.

The area which has still to respond and where much work is still needed is in the commercial, local authority and domestic sectors where fossil fuels are used for space heating, cooking and hot water supply only. Here the potential savings on fuel costs have not always been seen to be as significant as those in industry. Once the nettle has been grasped, a standard procedure of checks and balances can be adopted for an energy conservation programme.

A flow diagram best illustrates the strategies in the programme and is shown in Figure 7.4.

INITIAL COMMITMENT

As a result of potential future rising fuel costs and legislation on emissions of carbon dioxide, the attention of building owners will begin to turn toward the cost of fuel. The commitment will therefore eventually come from senior management in the organization. This is very important since the energy conservation programme will affect all those working in the organization and will cost money to implement.

Commitment to a programme of energy conservation currently is sluggish in the areas identified and needs the impetus of highly motivated and enlightened individuals.

APPOINTMENT OF THE ENERGY MANAGER

This may be an internal appointment or a consultant may be appointed from outside the organization. Either way, since energy conservation is about changing things and managing change, the individual must have authority and be non-partisan. The energy manager's responsibilities would be:

- initiating the energy conservation programme in the organization;
- undertaking an energy survey and energy audit on site;
- analysing the energy use on site and reporting to senior management;
- preparing cost and payback of identified conservation measures;
- implementing the agreed conservation measures;
- monitoring the results and comparing with forecasts;
- agreeing tariffs with the energy suppliers and purchasing the energy.

The responsibility of the energy purchaser for the organization might be seen as the first and only job that the energy manager need undertake. Judicious purchasing of energy will have the most immediate impact on fuel costs for the organization. However, purchasing is a one-off phenomenon and does not obviate the need for the prudent use of energy.

COLLECTING APPROPRIATE INFORMATION

Having obtained official support from senior management who must advise all the staff to assist in his or her endeavours, the energy manager

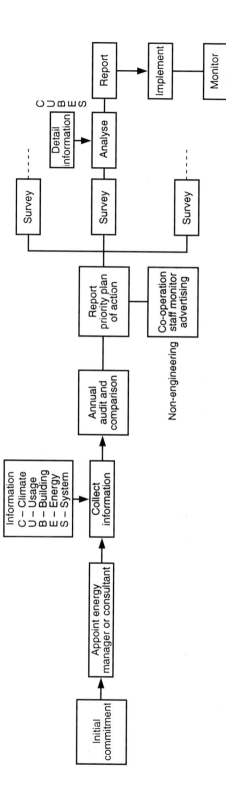

Figure 7.4 General flowchart of energy conservation programme [4].

then needs to collect all relevant information relating to the building's energy and use. This can be done under the following initials which relate to the flow diagram in Figure 7.4 [4]:

C   Climate: Degree Days for locality, mean daily outdoor temperatures, hours of daylight, hours of sunshine.
U   Usage: hours of occupancy, cleaning hours, holiday periods, special use, overtime periods, shifts, population, product output, water consumption.
B   Building: general state of repair of the external envelope, site plans, floor plans, elevations, U values, modifications, age, life.
E   Energy: fuel consumptions from fuel bills for coal, gas, oil, bottled gas, electricity, maximum demand values, tariffs, standing charges. Two to three years of data are required.
S   Systems: system layouts, design loads, manufacturers' outputs, nameplate data, modifications, controls schematic.

7.6 The energy audit

This is discussed in detail in Chapters 6 and 9. However, it is important to ascertain the extent and cost of energy use on the site. From the floor plans the floor area can be determined and the performance indicator for the building calculated in GJ/m^2 or kWh/m^2. An initial comparison with the SPIs for the building type can be made.

An energy audit allows the following matters to be aired:

• understand the scale of energy use and its annual cost;
• compare the energy use with a standard;
• make initial proposals on potential energy saving measures;
• budget cost the proposals;
• budget cost the potential savings on the fuel accounts.

7.7 The energy survey

The energy survey should result in an understanding of the energy flows within the buildings and identify energy wastage. A survey will have four principal objectives:

• to identify the points of energy input conversion, distribution and use;
• to identify the factors which are likely to affect the level of energy use for each area or item of plant;
• to identify areas of energy wastage and where efficiency could be improved by changing working practices, better maintenance, upgrading or replacement;
• to identify whether there is a programme of preventive maintenance for the building and its services.

Following the survey it should be possible to analyse the information and data from all sources and suggest means for improving efficiency and reducing wastage, and to calculate the level of potential savings for each case.

It may be that the instigation of a programme of preventive maintenance is the first task to be achieved by the energy manager. It is all too frequently not done properly. The three main activities in the energy survey are:

- locating the energy input and metering points, boiler plant, water heaters, etc.;
- energy conversion efficiency checks, responsibilities, etc.;
- energy utilization: tracking the converted energy, is it used efficiently, how is it controlled, responsibilities, etc.

Figure 7.5 shows a schematic diagram of a space heating system serving radiators, unit heaters and the heater battery in an air handling unit. Applying the three activities of the energy survey to the schematic diagram of the space heating system:

(i) Metering the fuel supply to the boilers, metering the power supply to the pumps, fans and controls.
(ii) Metering the heat supplied to the three circuits. If this is not possible because of lack of metering equipment, clamp-on thermometers can be used to find the flow and return temperatures at the circuit connections on the boiler header and at the points of use. Checking the boiler combustion efficiency test and its frequency.
(iii) Checking that the space heating appliances are operable. Are the areas in which the space heating appliances are located occupied when the heating circuit is on? Checking the frequency of cleaning to the unit heater and heater battery coils and the filter on the air handling unit. Checking the temperature controls and settings. Checking the drives on motors to pumps and fans. Checking the frequency of flushing out the heating system. Checking the thermal insulation to the distribution pipework.

You can see from the activities undertaken here that the four principal objectives in the survey can now be addressed. It is likely, however, that the only activity which comes out as positive in this survey is the meter reading on the fuel supply and that obtained from the fuel invoices which would have been read by the fuel supplier.

The energy manager may therefore have something to say at this stage in the survey report. In the absence of heat meters or flow meters, flow and return temperatures to each circuit can be checked using contact thermometers. See Appendix 5.

Some of the checks for the system shown in Figure 7.5 may appear to be basic. However, they are important and some should form part of the preventive maintenance programme for the services within the building anyway. If a building energy management system is installed

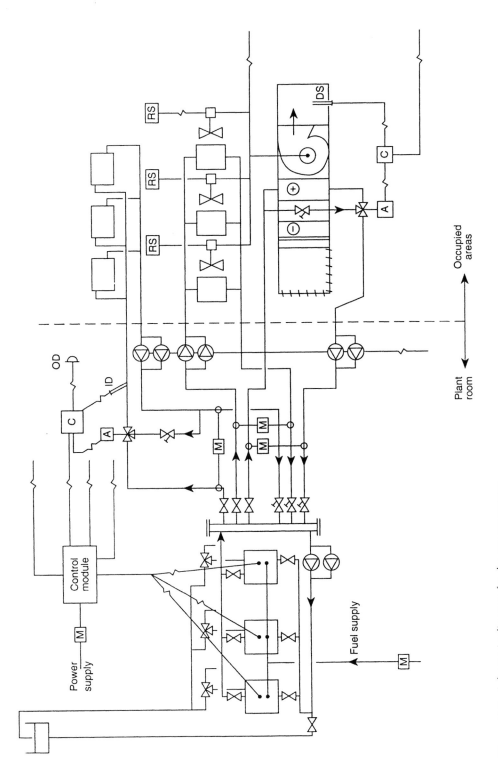

Figure 7.5 Schematic diagram for the energy survey.

some of the checks can be undertaken at the central station or outstation but a visit to each part of the building is essential.

7.8 Areas for energy saving

Refer to Appendix 2 for a comprehensive list. There are three areas which can be considered, namely, the building, the services within it and the usage requirements for the building.

1. **The building**
    Thermal insulation of the external envelope and reduction of unwanted infiltration of outdoor air are the two main factors. The latter will include the use of air locks at entrances, the use of self-closing doors in stairwells and on corridors, well-sealed windows and control of ventilation in lift shafts. The former will be helped by having a recognized programme of external maintenance.

2. **The services**
    There are two factors here: services provided for the occupiers and not requiring their intervention, like space heating, ventilation and air conditioning, and services used by the occupier, like hot and cold water supply, lighting, etc. Here are some areas for potential energy saving in the provision and use of services:
    - space heating: efficiency of generation, efficiency of distribution (pump power and thermal insulation), efficiency of utilization, time control, temperature control;
    - hot water services: temperature of supply, quantity of supply;
    - ventilation: quantity of fresh air, time control, temperature control, efficiency of distribution (fan power);
    - electrical power: control of use, correct tariffs and maximum demand;
    - air conditioning: free cooling, fan control, time, temperature and humidity control;
    - heat recovery: use and application of heat recovery devices such as thermal wheels, heat pipes, run around coils, plate heat exchangers, condensing boilers;
    - energy saving equipment: variable speed pumps and fans, weather compensated temperature controls, direct fired non-storage hot water supply plant, boiler water temperature controls.

    You should make yourself familiar with heat recovery and energy saving equipment available on the market.

3. **The usage requirements for the building**
    If time control for space heating is considered for a moment, an office having a nine-to-five day for five days a week over a 39-week

Example 7.2
The plant servicing a heating system with weather compensated temperature control providing constant volume variable temperature to circuits of radiators is to be replaced. The new plant will consist of one condensing boiler and one conventional boiler. The cost of replacement is estimated at £3000. From the data, determine the estimated saving in fuel and the simple payback period.

Data:
Annual cost of natural gas, excluding standing charge, £2800.
Efficiency of original boiler plant 70%.
Efficiency of new boiler plant 85%, made up of an efficiency for the new conventional boiler of 75% and that for the condensing boiler of 95%.
Charge for natural gas 1.3p/kWh.

Solution
Note that the efficiencies quoted are the heat conversion efficiencies provided by the boiler manufacturers. It is not necessary to consider seasonal efficiencies but it is important to compare like with like.

If each of the new boilers operates for 50% of the time, the average efficiency can be taken as $(95 + 75)/2 = 85\%$.
The annual energy consumption AEC $= 2800 \times 100/1.3 = 215\,385$ kWh.
The annual energy demand AED $= 215\,385 \times 0.7 = 150\,770$ kWh.
The new AEC $= 150\,770/0.85 = 177\,376$ kWh.
The estimated annual cost for gas $= 177\,376 \times 1.3/100 = £2306$.
The estimated annual saving $= £2800 - £2306 = £494$.
Simple payback $=$ capital cost of measure/annual saving.
The estimated simple payback $= 3000/494 = 6.1$ years.

Cost benefit analysis which includes simple payback is investigated in Chapter 8.

heating season requires heating for 1560 hours. The total time over that period is 6552 hours. The heating is therefore required for only 24% of the total time. The thermal insulation in the external envelope should therefore be close to the inner surface to keep the warm up time to a minimum, Mondays requiring the longest preheat. Plant run times are therefore important to energy conservation.

7.9 Heat recovery   One of the ways in which energy can be conserved is by the provision of heat recovery plant. As with any form of energy conservation the feasibility of introducing heat recovery equipment must be investigated. The following are key parameters:

- that there is a sufficient quantity of recoverable heat;
- the recovered heat can be made available at a suitable temperature;
- there is a use for the recovered heat;
- the waste heat source and point of reuse are not too remote;
- there is a match between the time of heat demand and the time of waste heat availability.

Some examples of successful heat recovery would include:

- the use of the recirculation duct on a ventilation/air conditioning system;
- the use of a plate heat exchanger in the return and fresh air intake ducts on a full fresh air ventilation/air conditioning system;
- heat recovery from the flue gases on space heating boiler plant;
- recovery of the sensible and latent heat from the return duct of a swimming pool ventilation plant.

Example 7.3

A tempered air mechanical ventilation system supplies 4.5 m³/s of air to a building located in London. Occupancy is five days per week, 12 hours per day. The ventilation plant is taken as responsive and the building is of medium weight. It is decided to undertake a feasibility study to consider introducing air recirculation in the air handling plant which currently operates on full fresh air.

Data:
Occupancy is 160.
Indoor design temperature is 19°C.
Outdoor design temperature is −2°C.
Minimum fresh air per person is 8 litres/s.
Air density is 1.2 kg/m³.
Specific heat capacity of air is 1.02 kJ/kg K.
Seasonal efficiency of boiler is 70%.
Cost of natural gas is 1.4p/kWh.
Capital cost of recirculation duct, mixing dampers and controls is estimated at £3500.

It is assumed that the pressure developed by the existing extract fan will cope with the altered ductwork.

Solution
From Table 1.5 the annual SDD is 2034.
From Tables 5.1, 5.2 and 5.3 the correction factors appropriate here, interpolated where necessary, are 0.8, 1.13, and 0.7 respectively.
Corrected DD = 2034 × 0.8 × 1.13 × 0.7 = 1287.
Equivalent hours of operation at full load = DD × 24/dt = 1287 × 24/(19 + 2) = 1471.
Minimum fresh air supply = 160 × 8/1000 = 1.28 m³/s.
Saving in design heat load = (4.5 − 1.28) × 1.2 × 1.02 × (19 + 2) = 83 kW.

Energy saved by recirculation in the heating season:

estimated annual energy saving = 83 × 1471 = 122 093 kWh;
estimated annual cost saving = (122 093/0.7) × 1.4/100 = £2442;
estimated simple payback = capital cost of measure/annual saving
$$= £3500/2442 = 1.43 \text{ years.}$$

Since the return on the investment is so short the scheme is financially feasible. It would then be necessary to consider the practical feasibility of the proposal although since the alterations have already been costed it is reasonable to assume that they are a practical possibility.

Example 7.4
An alternative proposal is now considered for the tempered air system serving the building in Example 7.3. One of the disadvantages of air systems incorporating recirculated air is that air contaminated with bacteria which cannot be cheaply filtered out is returned and inhaled by the occupants unless sophisticated and expensive air filtration is used. A continuous filtered fresh air supply is preferable and will assist in the elimination of the potential for sick building syndrome. The proposal is to consider the feasibility of using a plate heat exchanger as a recuperator between the fresh air supply and return ducts.

Data:
Air pressure drop in both the exhaust and fresh air sides of the recuperator is 190 Pa.
Heat transfer efficiency of the recuperator is 64%.
Fan efficiency is 68%.
Fan drive efficiency is 98%.
Fan motor efficiency is 92%.
Charge for electricity is 7.0p/kWh.
Capital cost of the recuperator including the extra costs to the supply and extract fans and motors in overcoming the increased resistance to air flow, insulation and controls is £6500.

Solution

The connections to the recuperator are shown in Figure 7.6. Note the bypass arrangement for use outside the heating season.

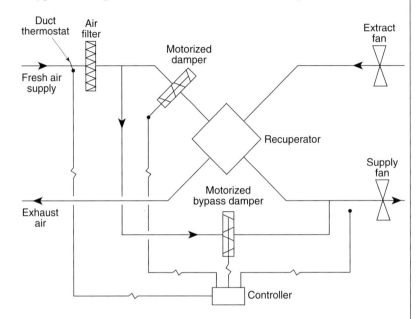

Figure 7.6 Application of a plate heat exchanger recuperator in supply and extract ducts.

Equivalent hours of operation at full load = 1471.
Saving in design heat load = 0.64 × 4.5 × 1.2 × 1.02 × (19 + 2) = 74 kW.
Estimated annual energy saving = 74 × 1471 = 108 854 kWh.
Estimated annual cost saving = (108 854/0.7) × 1.4/100 = £2177.
Fan power consumed in overcoming the air resistance in the recuperator is

$$2 \times 190 \times 4.5/(0.68 \times 1000) = 2.515 \text{ kW}.$$

Operating hours for the fans = 5 days/week × 12 hours/day × 39 weeks = 2340 hours.
Note: outside the heating season the recuperator is bypassed.
Electrical energy used = 2.515 × 2340/(0.98 × 0.92) = 6527 kWh.
Estimated annual charge = 6527 × 7/100 = £457.
Estimated net annual saving = £2117 − £457 = £1660.
Estimated simple payback = £6500/£1660 = 3.9 years.
Clearly the payback period is longer than that required for the provision of recirculated air. However, it provides a better quality of ventilation than the use of recirculated air.

In Example 7.4 only the sensible heat in the exhaust air is reclaimed. The enthalpy of a typical sample of room air might comprise 60% sensible heat and 40% latent heat.

If you were considering two heat recovery appliances, one operating on sensible heat reclaim and the other on total enthalpy and both had a stated efficiency of 70%, the latter would give a genuine 70% recovery but the former would recover only the sensible heat from the room air. Thus its efficiency would be 70% of 60% which is 42%.

Example 7.5
A swimming pool hall is supplied with full fresh air at 4.72 m³/s and 35°C db, 10% saturated. The outdoor air is sensibly heated from saturated conditions at −1°C. The design conditions for the pool hall are 28°C db and 70% saturated which are also the air conditions in the extract duct. Determine the sensible and latent heat content of the exhaust air and the rate at which moisture is being lifted from the pool. Determine also the design load on the air heater battery.

Solution
Figure 7.7 shows the winter cycle on a sketch of the psychrometric chart. You will need a copy of the CIBSE psychrometric chart to make full sense of the solution.

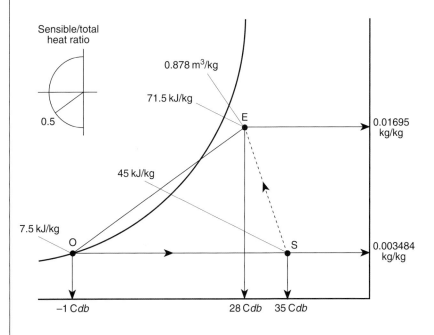

Figure 7.7 Psychrometric cycle without the thermal wheel: supply air at 35°C db, 10% saturated; exhaust air at 28°C db 70% saturated (Example 7.5).

The specific enthalpy of the air exhausted to atmosphere is 71.5 kJ/kg.

The specific enthalpy of the fresh air entering the air handling unit is 7.5 kJ/kg.

The heat content of the exhaust air $= M \times d_h = (v_{fr}/v) \times d_h$
$$= (4.72/0.878) \times (71.5 - 7.5)$$
$$= 344 \text{ kW}.$$

From Figure 7.7 the sensible to total heat ratio is about 0.5:1.0 and therefore the sensible heat content is $344/2 = 172$ kW and the latent heat content is 172 kW.

The moisture lifted from the pool $= (v_{fr}/v) \times d_g$
$$= (4.72/0.878) \times$$
$$(0.01695 - 0.003484)$$
$$= 0.07239 \text{ kg/s}$$
$$= 261 \text{ kg/hour}.$$

Clearly there is opportunity for heat recovery here, from both the sensible and latent heat from the exhaust air and the loss of treated and heated water from the pool.

If the pool is in use for 12 hours a day with the air handling plant operating, the water loss will be $261 \times 12 = 3132$ kg per day.

The design output of the air heater battery for the pool hall will be from point O to point S in Figure 7.7:

air heater battery output $= (v_{fr}/v) \times d_h = (4.72/0.878) \times$
$(45 - 7.5) = 202$ kW.

Substantial savings in energy demand and make-up water for the pool can be achieved with the use of a thermal wheel [5]. This is explored in Example 7.6.

Example 7.6

Consider the installation of a total enthalpy thermal wheel for the air handling plant serving the pool hall in Example 7.5 and identify the potential savings during a winter cycle. The manufacturers of the thermal wheel claim an efficiency of 70%.

Solution

Figure 7.8 shows a thermal wheel and Figure 7.9 shows the winter cycle on a sketch of the psychrometric chart. Note that the exhaust condition is now the condition of the air extracted from the pool hall before it is exhausted to atmosphere.

It is important to ensure that the moisture in the air does not condense out during the process of total heat recovery. A preheater is required, therefore, to sensibly raise the temperature of the incoming fresh air from the design saturated condition of $-1°C$ db to about 2.5°C db on wheel condition. This prevents the line on the chart between outdoor condition O and the extract condition from crossing the saturation curve.

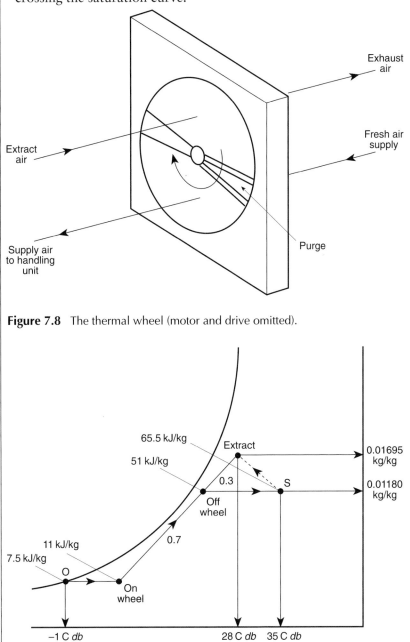

Figure 7.8 The thermal wheel (motor and drive omitted).

Figure 7.9 Psychrometric cycle with the thermal wheel at 70% efficiency (Example 7.6).

The off wheel condition will be 7/10 along the line on wheel to the extract. An after heater is needed here to sensibly heat to supply condition S. The combined outputs of the preheater and after heater will be less than the air heater battery required in Example 7.5 so there is already a saving here.

The heat content of the exhaust air was = 344 kW.

Referring to Figure 7.9:
the energy saved by the total enthalpy thermal wheel is

$$(4.72/0.879) \times (51 - 11) = 215 \text{ kW};$$

the heat content of the exhaust air will now be

$$344 - 215 = 129 \text{ kW}.$$

This represents a heat recovery from the exhaust air of 215/344 = 62.5%.
The design load on the preheater will be

$$(4.72/0.878) \times (11 - 7.5) = 19 \text{ kW}.$$

The design load on the after heater will be

$$(4.72/0.878) \times (65.5 - 51) = 78 \text{ kW}.$$

The total sensible heating load at design conditions is

$$19 + 78 = 97 \text{ kW}.$$

This compares with the design load of 202 kW in Example 7.5 before the use of the thermal wheel. This represents a potential saving on sensible heating of (202 − 97)/202 = 52%.
The moisture lifted from the pool is also reduced offering a further saving:

$$\text{moisture lifted} = (4.72/0.878) \times (0.01695 - 0.01180)$$
$$= 0.02769 \text{ kg/s}$$
$$= 100 \text{ kg/hour}.$$

Comparing with Example 7.5 for a 12-hour period the moisture lifted from the pool is now 1200 kg/day instead of 3132 kg. This represents a saving in make-up water, water treatment and heating for the pool of 62%.

SUMMARIZING EXAMPLES 7.5 AND 7.6

The design data before and after the proposed use of the total enthalpy thermal wheel is tabulated in Table 7.2.

Table 7.2 Summary of solutions to Examples 7.5 and 7.6

	Before	After	Saving
Rate of heat discharge in exhaust	344 kW	129 kW	62.5%
Design sensible heat load	202 kW	97 kW	52%
Water loss from pool	3132 litres/day	1200 litres/day	62%
Make up pool water treatment	3132 litres/day	1200 litres/day	62%
Make up pool water heating	3132 litres/day	1200 litres/day	62%

When you check the solutions to Examples 7.5 and 7.6 you may find small discrepancies in reading off the specific enthalpy values from the psychrometric chart.

The savings indicated will be achieved only when the outdoor design condition is saturated air at −1°C db. It would be reasonable to assume that these therefore represent maximum savings. A more realistic approach would be to determine the savings from an average outdoor winter condition for the locality.

The use of the total enthalpy thermal wheel begs serious consideration, however.

Only the winter cycle has been analysed here and it would be appropriate to consider the operation of the plant during the summer. You should now consider the summer cycle taking outdoor air at say 28°C db, 70% saturated. Is additional air handling plant required? If you are in doubt you should seek advice. Alternatively recourse can be made to another book in this series [6].

Consider now the alterations required for the installation of the thermal wheel.

It would be necessary to have the fresh air intake and exhaust ducts located adjacently at a point upstream of the proposed preheater so that the thermal wheel can be positioned astride the two air streams. See Figure 7.8. This and the replacement of the air heater battery and the increased resistance to air flow will require substantial alterations to be made to the air handling plant with bypass ducts around the thermal wheel.

It is now suggested that you consider the potential energy savings

Table 7.3 Summary of solution using an outdoor condition of 6.5°C db, 80% saturation

	Before	After	Saving
Rate of heat discharge in exhaust	284 kW	81 kW	71%
Design sensible heat load	150 kW	73 kW	51%
Water loss from pool	2817 kg/day	871 kg/day	69%
Make up water treatment	2817 kg/day	871 kg/day	69%
Make up water heating	2817 kg/day	871 kg/day	69%

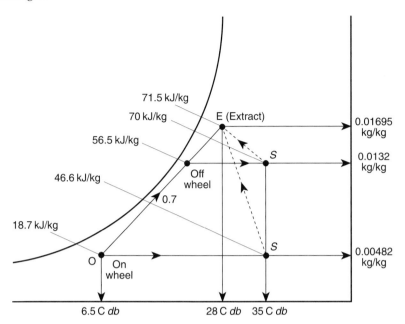

Figure 7.10 Psychrometric cycles with and without the thermal wheel; outdoor condition 6.5°C, db, 80% saturated

using an average outdoor winter condition (for the Thames Valley) of 6.5°C db, 80% saturated. The solution is tabulated in Table 7.3 and the cycle is shown on a sketch of the psychrometric chart in Figure 7.10.

Note: for an average outdoor condition of 6.5°C db no preheating is required. The preheater must be in place however to satisfy the requirements at the design outdoor condition of saturated air at −1°C db. Contrary to the initial assessment, the savings look potentially better than for operation at the design outdoor condition.

7.10 Chapter closure

You now have the tools to proceed with an energy survey which is underpinned with the strategies, checks and balances accounted for in the chapter. Some work has been done on the application of heat recovery equipment. You will have seen, therefore, how important it is to be knowledgeable in the design of heating, ventilating and air conditioning systems. Recourse can be made, relating to system design, to other publications in this series [6, 7].

Cost benefit analysis 8

Nomenclature

CPV   cumulative present value factor
DF   discount factor
n period of term in years
NPV   net present value
P principal sum
PV   present value
PVA   present value of an annuity
PWF   present worth factor
R return on investment
r rate of interest
SPB   simple payback
TV   terminal value
TVF   terminal value factor

8.1 Introduction

It is a fact of life that much of the way modern society organizes itself requires accountability. This, it might be argued, is a fine sentiment because it provides a check and a balance upon error and excess. In the current market place economy the requirement of accountability means cost accounting.

The quality of the environment in which we live is dependent upon scientific evidence and the pressure which is put on governments to prioritize issues relating to the environment and thus upon us members of society as to how much we are willing to pay for cleaner air, reduction in global warming, treatment of waste, reduction in the production of pollutants in manufacturing processes, and so on. The matters which concern us here are those of operating costs and energy management which happen to impinge on these same issues: the cleanliness of the air, global warming and the production of pollutants.

However, the energy manager or consultant has to show senior management or the client that the costs of saving energy (which will contribute, even if it is in a very small way, to an increase in the quality of the environment) are met by cost savings in the consumption of fossil fuel and hence a reduction in the expenditure column of the company's annual balance sheet. We have thus returned full circle to cost accounting or cost benefit.

There are various ways of looking at the benefits of cost accounting measures for energy saving. In some of the examples so far in the text simple payback is used. This is acceptable if inflation and interest rates are low and the period of payback is short. Some organizations prefer simple payback and simple costing of alternative energy saving schemes because of the vagaries of future inflation and interest rates.

However, the client or investor may want to compare the potential return on an investment in a proposal for energy saving with the potential return on investment in another aspect of the business. For this reason other ways of costing the benefits attributable to energy saving proposals are considered in this chapter and compared with simple payback.

8.2 Simple payback

This is defined as

SPB = cost of energy saving proposal/annual saving – annual cost of saving.

The annual cost of the energy saving proposal might, for example, be the cost of electricity for operating the energy saving equipment and/or the annual cost of maintenance on the energy saving proposal. Have a look at the examples and solutions in the previous chapters in which simple payback has been calculated.

LENGTH OF THE PAYBACK PERIOD

Payback periods of up to five years are usually acceptable. It is a current fact of life that longer periods of payback are not usually considered, much in the same way as long term investment, commercial, public and private, does not immediately attract the potential investor. Clearly the payback period must be less than the life of the energy saving measure and less than the life of the building and the services installation as a whole.

8.3 Discounted cash flow and present value

Consider an investment of £100 at an annual interest rate r (rate of return or discount rate) of 5%.

After the first year the sum would be $100 \times 1.05 = £105$.
After the second year the sum would be $100 \times (1.05)^2 = £110.25$.
After n years the original sum P would be the
terminal value $TV = P(1 + r/100)^n$

Alternatively we can say that the present value PV of £110.25 at the end of year 2 is worth £100 today if discounted at 5%. Thus

$$PV = P(1 + r/100)^{-n}.$$

Substituting, we have

$$PV = 110.25(1 + 5/100)^{-2} = £100.$$

The following case study looks at discounted cash flow.

Case study 8.1

Consider a situation where £3000 is spent out of an organization's reserves on an energy saving measure which has no further cost and the annual savings in fuel are £1000. The simple payback period will therefore be 3000/1000 = 3 years. Undertake a discounted cash flow and present value analysis and compare the results.

SOLUTION

If the money spent on the energy saving measure had instead been invested for three years at a discount rate of 5%, the terminal value would be:

$$TV = 3000(1 + 5/100)^3 = £3473$$

and the return R on the investment would be:

$$R = 3647 - 3000 = £473.$$

The return on the investment in the energy saving measure at the end of the fourth year will be £1000 net and £1000 net annually thereafter for the life of the measure, as long as fuel prices remain in line with general inflation. There are a number of points which can be made here:

- There is clearly no return on the investment in the energy saving measure during the first three years as the annual savings are paying for the original investment.
- The charge for servicing the loan of £3000 must be accounted for as well.
- If the life of the energy saving provision is 20 years, the net savings will accrue for the remaining 17 years and amount to £17 000 at present value.
- If the energy saving measure had not been undertaken and the £3000 left in the organization's reserves, after 20 years at 5% it would be worth £7960 providing a return of £4960.

There is clearly a strong motive to prosecute the energy saving provision.

• The net savings accruing from the energy conservation measure are £17 000 – 4960 = £12 040.

An alternative approach is normally adopted in which the payback period accounts for the initial loss in earnings. The results of the energy saving provision are given in Table 8.1.

Note: present value factor $= 1/(1 + r/100)^n = (1 + r/100)^{-n}$ (8.1)

PV = energy saving × PV factor.

A table of PV factors is to be found in Appendix 6 for ascending values of r and n.

Some sources employ the discount factor and DF $= (1 + r/100)^n$ in which case the present value is obtained from PV = energy saving/discount factor.

From the results in Table 8.1 we see in the cumulative PV column that although simple payback showed that the initial outlay would be paid for in three years, the effect of the interest rate means that it takes between three and four years before the energy saving measure is paid for.

It is possible to determine accurately what the payback period will be from the following formula:

Cumulative present value factor

$$CPV = (1 - (1 + r/100)^{-n})/(r/100). \qquad (8.2)$$

The cumulative present value factor is also known as the present worth factor (PWF) and present value of an annuity (PVA). A table of PVA factors is given in Appendix 6 for ascending values of r and n.

Also,

CPV = simple payback period.

Therefore in this case study:

$$3 = (1 - (1 + 5/100)^{-n})/(5/100)$$

Table 8.1 Tabulated solution to case study 8.1

Start of year	Cost of measure	Energy saving	PV factor	PV	Cumulative PVs
0	£3000	–	–	–£3000	–£3000
1	–	£1000	0.953	+£935	–£2065
2	–	£1000	0.907	+£907	–£1158
3	–	£1000	0.864	+£864	–£294
4	–	£1000	0.823	+£823	+£529

from which:

$$0.15 = 1 - (1.05)^{-n}$$

and

$$(1.05)^{-n} = 0.85$$

then

$$(1.05)^{n} = 1.176.$$

Taking logs to the base 10:

$$n \log 1.05 = \log 1.176.$$

Therefore

$$n = \log 1.176/\log 1.05$$

and the payback period $n = 3.331$ years.

INITIAL SUMMARY TO CASE STUDY 8.1

- Simple payback takes three years for the original cost of the energy saving provision to be paid.
- Discounted payback takes 3.33 years for the original investment to be paid. This accounts for the loss in revenue which would otherwise accrue by leaving the £3000 invested at an interest rate of 5%.
- Now have a look at case study 2.1 in which CPV is used.

It is likely that to reduce levels of carbon dioxide emissions building owners will need an incentive to concentrate the mind on how energy consumption can be reduced in their building stock. There are at least two ways of achieving this. Tax incentives can be introduced or the government of the day can institute a progressive increase in the value added tax on fossil fuels.

   Taking case study 8.1 and working on the assumption of nil general

8.4 Effects of fuel inflation (case study 8.1)

Table 8.2 Accounting for fuel inflation in case study 8.1

Start of year	Cost of plant	Energy saving	Energy inflation	Cash flow	PV factor	PV	Cumulative PV
0	−£3000	–	–	−£3000	1.0	−£3000	−£3000
1	–	£1000	1.06	+£1060	0.952	+£1009	−£1991
2	–	£1000	1.06^2	+£1124	0.907	+£1019	−£972
3	–	£1000	1.06^3	+£1191	0.864	+£1029	+£57

inflation but an annual inflation on fossil fuel of 6%, the following results will occur as shown in Table 8.2.

Note: cash flow = energy saving × annual energy inflation.

Annual energy inflation is obtained from the terminal value factor:

$$\text{TVF} = (1 + r/100)^n. \qquad (8.3)$$

A table of terminal value factors is given in Appendix 6 for ascending values of r and n.

It can be seen from the cumulative present value column in the tabulated results that the measure will pay for itself in about 2.9 years.

FINAL SUMMARY OF CASE STUDY 8.1

Capital cost of measure	Annual saving	Payback period	Method of calculation
£3000	£1000	3 years	simple payback
£3000	£1000	3.33 years	5% discount
£3000	£1000	about 2.9 years	5% discount and 6% annual fuel inflation

As one would expect, progressive inflation of fuel prices gives a better return on the capital investment in energy saving measures by reducing the payback period and hence provides the incentive on the part of the building owner. The difficulty lies in assessing what the future fuel inflation and interest rates will be.

Three formulae have been introduced in this section:

$$\text{PVF} = (1 + r/100)^{-n} \text{ (equation (8.1))}.$$

This formula accounts for depreciation or inflation.

$$\text{CPV} = (1 - (1 + r/100)^{-n})/(r/100) \text{ (equation (8.2))}.$$

This formula enables annual repayments on a loan to be calculated. It also enables the calculation of the present value of annual costs over a fixed term to be done.

$$\text{TVF} = (1 + r/100)^n \text{ (equation (8.3))}.$$

This formula allows the calculation of compound interest on a single sum invested over a fixed term to be carried out. It is also used here to calculate the annual energy inflation on fuel.

Tables of PV factor, CPV and TVF are to be found in Appendix 6 for ascending values of r and n.

There is frequently a requirement to compare two or more proposals for the purposes of making a decision relating to the most financially viable project. An example of this is found in the following case study.

Case study 8.2

An oil fired boiler plant has come to the end of its useful life and a comparison is to be made between the installation of a new oil fired boiler plant and a new gas fired plant. After flushing out and replacing pumps and controls, the existing space heating system is otherwise satisfactory and will have an increased life of at least 20 years which will be the life of the new boiler plant. As work to the rest of the space heating system and removal of the existing boiler plant is similar for both proposals it does not need consideration in the comparison.

The information in Table 8.3 relates to each of the two proposals.

SOLUTION

A discount rate of 4% will be applied for case study 8.2. The factor to be applied for recurring annual expenditure is calculated from equation (8.2) or from the appropriate table in Appendix 6. The life of the system is taken as 20 years ($n = 20$), taking $r = 4\%$. Thus from Appendix 6 or equation (8.2) CPV = PVA = 13.59.

Table 8.3 Data for case study 8.2

Capital costs	Oil fired plant	Gas fired plant	Remarks
Refurbishment to chimney	£1200	£4900	Diluted flue for gas
Boiler plant	£12000	£9000	
Gas main	–	£1600	
Energy costs	£6000/yr	£4800/yr	
Maintenance and operating costs	£1000/yr	£800/yr	
Repair costs:			
3-year	£650	£370	
8-year	£950	£450	
12-year	£750	£400	
16-year	£1000	£750	
Repainting oil storage tank and bund	£750	–	

Table 8.4 Analysis of oil fired option, case study 8.2

Item	Expenditure	Factor	PV
Chimney	£1200	–	£1200
Plant	£12000	–	£12000
Energy cost	£6000/yr	13.59	£81540
Maintenance	£1000/yr	13.59	£13590
Repair:			
3 years	£650	0.889	£578
8 years	£950	0.7307	£694
12 years	£750	0.6246	£468
16 years	£1000	0.5339	£534
Repaint: 12 years	£750	0.6246	£468
		NPV = £111 072	

The cost of repairs at set intervals are single items of expenditure and the factor to be applied is calculated from equation (8.1) or obtained from the present value table in Appendix 6.

When

$n = 3$, PV factor $= 0.889$
$n = 8$, PV factor $= 0.7307$
$n = 12$, PV factor $= 0.6246$
$n = 16$, PV factor $= 0.5339$.

The solution to the oil fired option is given in Table 8.4. The solution to the gas fired option is given in Table 8.5.

Table 8.5 Analysis of gas fired option, case study 8.2

Item	Expenditure	Factor	PV
Chimney	£4900	–	£4900
Plant	£9000	–	£9000
Gas main	£1600	–	£1600
Energy cost	£4800/yr	13.59	£65232
Maintenance	£800/yr	13.59	£10872
Repair:			
3 years	£370	0.889	£329
8 years	£450	0.7307	£329
12 years	£400	0.6246	£250
16 years	£750	0.5339	£400
		NPV = £92 912	

CONCLUSION

You will notice here that we are interested in the present cost or net present value. The capital outlay in each proposal is a present cost and therefore has no factor applied.

From the tabulated calculations for the two proposals it is clear from the net present value totals that the gas installation is more economical. However, the capital outlay is less for the oil fired plant: for gas it is £15 500 and for oil it is £13 200. There is not much difference here in capital outlay but the client would want to know this information as well as the NPV costs.

8.6 Loans

If a building owner has to borrow money to service an energy saving measure the rate of interest chargeable on the loan must be accounted for in the financial appraisal. The interest charged on the loan will be greater than the interest given on a capital investment of the same value. Some building societies offer interest rates on investments at a lower rate than they charge on loans in order to have working capital to pay their employees. Banks have rates of interest for investors lower than the borrowing rate in order to pay their employees and make a profit.

In case study 8.1 the money raised for the energy saving measure was taken from the organization's reserves which were invested at 5%. If the money had instead been borrowed at an interest rate of 10% the payback period would be extended from 3.33 years to 3.74 years. You should now adopt equation (8.2) and check this solution for yourself.

The same is true in case study 8.2, thus affecting the calculation of the present value factor PV and the cumulative present value factor CPV.

There are a number of ways in which loan repayments are calculated as shown in case study 8.3.

Case study 8.3

A loan of £3000 is charged at a borrowing rate of 8% annually over a period of five years.

(a) Consider a loan the repayments of which are based upon interest on the capital sum

$$\text{total repayment} = £3000 + (3000 \times 0.08 \times 5)$$
$$= £4200.$$

Annual repayment $= 4200/5 = £840$
Monthly repayment $= 840/12 = £70.$

(b) Consider a loan, the repayments of which are based upon the annual cost method. From the table of CPV factors in Appendix 6 or the use of equation (8.2),

CPV = 3.993.

The annual cost for repayment of the loan = loan/CPV = 3000/3.993 = £751.
The monthly premium will be 751/12 = £62.6.
The total cost of the loan will be = 751 × 5 = £3755.

(c) Consider a loan based upon compound interest:

$$TV = 3000(1 + 8/100)^5$$
$$= £4408.$$

Annual repayment = 4408/5 = £882
Monthly repayment = 882/12 = £73.5.

SOLUTION SUMMARY

Repayment scheme	Monthly premium	Loan charge
(a)	£70	£1200
(b)	£62.6	£755
(c)	£73.5	£1408

You can see that the repayment or servicing of the loan varies in value and depends upon the method adopted for the calculation. The annual cost method (b) is normally used for undertaking cost benefit analysis.

8.7 Life cycle costs

The cost of owning a system over its life plus the cost in use establishes the life cycle cost. The life of various plant and systems is tabulated in the *CIBSE Guide* [1] and is usually between 15 and 30 years. Clearly the life of sophisticated plant will be less than that of benign systems such as cast iron radiator heating which will extend well beyond 30 years if properly maintained. It therefore may be necessary to consider the life cycle of a system in two parts, namely plant and distribution. Life cycle costs include the cost of the capital outlay, taking account of the interest it could have earned had it been invested, or the interest that must be paid on it had it been borrowed. The following case study illustrates life cycle costing.

Case study 8.4

The capital cost for a building services installation is £145/m². It is estimated to have a useful life of 20 years. The annual cost in use which includes heating, electricity and maintenance are estimated to be £20/m². Determine the net present value of the installation and the annual cost of owning and operating it.

If the project is financed out of the organization's profits or reserves a discount rate of 4% is to be used. If, alternatively, the capital has to be obtained in the form of a loan from a bank, a borrowing rate of 7% is to be adopted. Ignore the effects of inflation.

SOLUTION

Using the table of cumulative present value factors in Appendix 6 or equation (8.2):

For a discount rate of 4%, CPV = 13.59.
The present value of the capital cost is £145.
The present value of the cost in use is £20 × 13.59 = £272.
Therefore total NPV = £417/m².

For a discount rate of 7%, CPV = 10.594.
The present value of the capital cost is £145.
The present value of the cost in use is £20 × 10.594 = £212.
Therefore total NPV = £357/m².

To repay a loan of £145/m² over 20 years at a borrowing rate of 7% using the annual cost method, the CPV from the table in Appendix 6 or from adopting equation (8.2) is 10.594.

Loan on first cost payable each year = loan/CPV = 145/10.594
= £13.69.
Annual cost in use is estimated at £20.
Total annual owning and operating cost = £33.69/m².

If the building owner took capital from company profits or reserves to pay for the installation the loss in interest on this capital which would otherwise be invested would be 4%, and adopting the annual cost method of repayment, the CPV over a term of 20 years would be 13.59.

Loan on first cost payable each year = loan/CPV = 145/13.59
= £10.67.

Annual cost in use is estimated at £20.
Total annual owning and operating cost = £30.7/m².

SUMMARY OF LIFE CYCLE COSTS FOR CASE STUDY 8.4

Net present value of the installation: NPV	=	£357/m² borrowed capital.
Total annual owning and operating cost	=	£33.7/m² borrowed capital.
Net present value of the installation: NPV	=	£417/m² investment from capital or reserves.
Total annual owning and operating cost	=	£30.7/m² investment from capital or reserves.

• Note that the net present value of the total annual owning and operating cost of £33.69/m² on the same terms, is 33.69 × CPV, which is 33.69 × 10.594, and NPV = £357/m².
• Likewise the NPV of £30.7/m² = 30.7 × 13.59 = £417/m².
• Note also that the NPV is lower for the higher borrowing rate (7%) than it is if investment is taken from capital or reserves (4%).
• The annual owning and operating cost is however lower if capital or reserves are used to finance the project.
• Allowances on taxation of company profits may be available on capital expended on investment in the business. This will have the effect of reducing further the total annual owning and operating costs if the organization finances the project from reserves.

Another example of life cycle costs is the comparison of alternative schemes illustrated in case study 8.2.

8.8 Repair or replace

When an installation is approaching the end of its economic life, consideration should be given as to whether to refurbish it or replace it. It is likely that in a number of cases partial replacement will be chosen since parts of the installation may have a longer life. For example, the distribution pipework and radiators in a space heating installation. The factors which require consideration are:

(i)   cost of replacement;
(ii) energy saving benefit of new plant;
(iii) cost in use of new versus existing plant;
(iv) value placed on reliability and safety if existing plant is kept.

Factor (ii) forms part of factor (iii) but is separately listed to emphasize the opportunity that new and more efficient plant will have in saving energy and in reducing carbon dioxide emissions.

    The criterion to be followed in making a decision is to compare the NPV of both options. This will include comparing the cost of new replacement plant and its cost in use, which will reflect the savings in energy consumption, with the repair costs of existing plant and its cost in use. The analysis can be presented as that in case study 8.2.

You now have some insight into cost benefit analysis which includes simple payback, discounted cash flow and present value, effects of fuel inflation, an NPV comparison of alternative schemes, calculation of premiums on loans and life cycle costs. The important focus here is to set out the energy saving proposal or comparison of alternative schemes or proposals in such a way that they are meaningful to the company accountant, so that senior management can make financial decisions relating to energy saving proposals alongside other financial matters.

8.9 Chapter closure

9   Energy audits

Nomenclature

AEC   annual energy consumption
AED   annual energy demand
AHU   air handling unit
ECY   energy consumption yardsticks
HWS   hot water supply
LPG   low pressure gas
PI   performance indicator

9.1 Introduction

A definition of the term 'energy audit' may help to put this chapter into context. An energy audit attempts to allocate a value at each point of energy consumption over a given period, usually a year. It should at the least allocate a value to the consumption of the various forms of energy on site. Following an energy audit it should then be possible to identify at each point of energy consumption the corresponding energy demand. This may require an analysis of energy demand at each point. The difference between energy consumption and energy demand can then be addressed and ways considered to reduce it at each point to its smallest practical value.

In Chapter 7 the relationship between annual energy consumption, AEC, and annual energy demand, AED, was identified as:

$$AEC = AED/(seasonal)\ efficiency, GJ\ or\ kWh.$$

9.2 Preliminaries to an energy audit

Before proceeding with an energy audit it is essential that a survey of the staff who use the buildings is undertaken to establish if the comfort levels are acceptable during the winter season and to find out if the provision of hot water for consumption is adequate. It is not uncommon in a survey of this kind to find that complaints arise due to discomfort from the lack of adequate space heating. This may be due to a variety of reasons. The effect this will have on the energy audit may well be significant, as it will show a false annual energy consumption. If the standards of comfort in those areas of complaint are addressed after the audit there will be an increase in annual energy consumption in the following year. This will give a negative signal to the client or senior management.

If complaints of overheating are found or if the workplace is said to be too cold, as evidenced by open windows on cold days, this could be the first issue to be addressed in the following programme of energy conservation. For sedentary occupations in modern well-insulated buildings, the majority of people will be neither uncomfortably cool nor uncomfortably warm in winter in rooms where the dry resultant temperature is between 19 and 23°C. For active occupations in modern well-insulated buildings, the dry resultant temperature should be between 3 and 5°C below that for work of a sedentary nature. Recourse should be made to Energy Efficiency Office's *Fuel Efficiency booklets* [1] which are currently supplied free of charge.

9.3 Outcomes of the energy audit

Clearly if the efficiency of energy use can be improved the difference between AED and AEC will be reduced. The efficiency of energy use in the case of boiler plant for space heating over a period will be its seasonal efficiency.

Modern boiler plant and controls are more efficient at converting primary energy than older plant. There have been significant advances in boiler and burner design. New plant like the condensing boiler which has a very high efficiency has entered the market in recent years. The concept of connecting low output boilers in modular format to match changes in load is energy efficient. The move toward generating hot water supply in direct fired instantaneous heaters is energy efficient. The move toward decentralization of boiler plant to remove energy losses in distribution is energy efficient, although this should not be confused with group heating and district heating schemes which can make use of refuse incineration, power generation and waste heat.

There are, however, other ways in which annual energy consumption can be reduced and these involve an equally valid approach, namely to reduce the annual demand for energy. By reducing AED, AEC should also be reduced. There are a variety of action points here and they include:

- checking maintenance schedules and certificates to ensure that plant and system preventative maintenance has been undertaken, see Appendix 3;
- checking the time scheduling of plant and zones;
- checking thermostat settings;
- checking the thermal insulation on distribution mains;
- identifying and costing potential energy savings from improvements in the thermal insulation of the building;
- identifying and costing reduction of the ingress of unnecessary

outdoor air into entrance areas and corridors via entrance doors, stairwells and lift wells and around widows;

- encouraging the occupants of the building to view energy conservation as a daily habit.

The last action point is perhaps the most important. The education of staff in an awareness of energy conservation is paramount to the success of an energy conservation initiative. The energy audit does not of itself reduce energy use on a site. It can, however, assist in initiating action to reduce the annual energy cost. The energy audit therefore provides two potential courses of action:

- to improve the efficiency of energy conversion to reduce the difference between AED and AEC to its lowest practical value;
- to reduce AED to its lowest practical value.

9.4 Measurement of primary energy consumption

In order to prepare an energy audit the invoices for all primary energy supplied to the site must be available for at least the penultimate calendar year. Adjustments may have to be made to ensure that the invoices for different fuels cover the same period. The use of estimated readings should be avoided or checked. At the same time the client should be encouraged to have the various fuels used on site accurately metered so that the fuel accounts for the following year are precise.

Measurement of the consumption of electricity, gas and bulk fuels such as petrol, diesel oil, heating oil, coal, bottled propane and butane, can be monitored at the incoming point on site for each. However, the client should be advised to meter the fuels and energy consumption at the various points of use so that in time a more detailed energy audit at points of use can be undertaken. This type of audit identifies annual energy consumption at points of use and it is from this information that further conservation studies emanate. Refer to Appendix 5 for details of monitoring equipment.

9.5 Primary energy tariffs

One of the energy manager's responsibilities is the negotiation of tariffs with the privatized utilities. It is not in the scope of this book to provide advice in this area which has now become quite specialized. There is no doubt, however, that negotiation of competitive tariffs will have a bearing upon the annual cost of energy, but it is a one-off procedure – at least until the tariffs come up for renegotiation. It is therefore only one of many potential energy conservation measures. Energy consumption features to take into account when investigating the energy supply markets are:

- annual base load requirement;
- peak loads and their frequency and points of occurrence;

- penalties for exceeding peak loads (and ways this can be avoided by load shedding);
- seasonal load;
- security of supply (and the potential need for back-up supplies);
- in the consumption of electricity, power factor correction.

You can see that these features require detailed knowledge of the way primary energy is consumed on the site. This data may not be fully available from historical records and the energy manager may need to interview key occupants/employees for evidence at points of energy use.

The initial annual energy audit can be presented in tabulated format as shown in Table 9.1.

9.6 Presenting the data – a simple audit

CONVERSION OF ENERGY FROM FUELS TO A COMMON BASE

Energy is normally expressed in GJ or kWh. Table 9.2 gives conversions for the common fuels.

There now follows an example of an energy audit based upon historical fuel invoices.

Case study 9.1

Fuel consumption and cost data are drawn from the previous year's accounts of an organization and detailed below. Prepare an initial energy audit and comment upon the results.

Table 9.1 A format for an initial energy audit

1. Fuel	Gas	Oil	Electricity	LPG	Totals
2. Consumption (litres, kWh, bottles)					
3. Consumption in common units					Total
4. Annual cost (£)					Total
5. Cost in GJ, kWh					
6. Percentage of total cost					100%
7. Percentage of total consumption					100%

Line (1) of the table should show all types of energy used on the site; four examples are shown here.
In line (2) consumption should be based upon fuel invoices and best estimates where necessary.
Line (3) shows consumption figures converted to a common unit, either GJ or kWh.
Line (4) shows the total annual cost of the energy/fuel inclusive of standing charges where appropriate.
Line (5) = item in line (4)/item in line (3).
Line (6) = item in line (4)/total cost in line (4).
Line (7) = item in line (3)/total cost in line (3).

Table 9.2 Energy conversions for common fuels

Fuel	Energy produced	
	GJ	kWh
1 tonne of coal (average)	27.4	7617
1 litre of petrol	0.04	11.12
1 cubic metre of natural gas	0.0387	10.76
1 cubic foot of natural gas	0.0011	0.305
1 therm of natural gas	0.1055	29.33
1 cubic metre of propane gas	0.0926	25.74
1 kg of propane gas	0.05	13.9
1 litre of light fuel oil	0.0405	11.26
1 litre of heavy fuel oil	0.0412	11.45
1 kWh of electricity	0.0036	1.0

Gas: 12 760 therms at £4225; light grade oil: 8000 litres at £960; electricity: 110 000 kWh at £4600; propane: 1600 kg at £480.

SOLUTION

Clearly much work on the part of the energy manager dealing with this site has been completed already in the production of the fuel account data since it will not be readily available. It is likely that key staff were unaware that the organization was buying in four different forms of energy. It is also possible that senior management would not immediately know that it spent £10 265 in the previous year on energy.

The results are given in Table 9.3.

Table 9.3 Solution to case study 9.1

	Fuel and consumption				
	Gas, 12 760 therms	Oil, 8000 litres	Electricity, 110 000 kWh	Propane, 1600 kg	Totals
Consumption, common units GJ	1346	324	396	80	2146
Annual cost	£4225	£960	£4600	£480	£10 265
Cost/GJ	£3.14	£2.96	£11.62	£6.00	
Percentage of total cost	41%	9%	45%	5%	100%
Percentage of total consumption	63%	15%	18%	4%	100%

COMMENTS ON RESULTS OF CASE STUDY 9.1

- The total cost of energy for the previous year is now known.
- As would be anticipated the major fuel is gas at 63%.
- The most expensive 'fuel' is electricity, making up 45% of the annual energy cost at only 18% of annual energy consumed.
- Unless there is obvious potential for saving energy in the use of the other fuels, it is clear that gas and electricity consumption should be investigated first, since the major fuel consumption will offer the greatest scope for energy saving and the most expensive fuel will offer the greatest cost benefit for every unit of energy saved.
- At present the cost of oil corresponds to that for gas and if the boiler plant is coming to the end of its working life it may be worth investigating the market for high efficiency dual fuel plant to take advantage of a change in tariffs. However, oil has to be stored on site and this facility may take up valuable space and therefore generate a capital and ongoing cost.
- It is important, having produced the energy audit, to identify an immediate potential cost benefit which could result from an energy saving measure. This will help to justify the cost of the energy manager's services. If the organization is typical of most and has not previously undertaken an energy audit, it is most likely that a low cost energy saving measure like restructuring the fuel tariffs would bring a cost benefit with no capital outlay.
- If the treated floor area of the premises is calculated, performance indicators can be determined and compared with standard performance indicators. See Chapter 6.

The next case study is in two parts. The first looks at an initial audit evaluated from invoices for gas meter and electricity meter readings for a site. The second audit is determined from more detailed observations on the same site for the same calendar year. It will show the value of investing in metering equipment.

Case study 9.2

A building consisting of five storeys and having a treated floor area of 2000 m² is divided into offices, restaurant, gymnasium and aerobics room. The quarterly gas and monthly electricity consumptions were taken from the main utility meters and are given in Table 9.4.

SOLUTION

You will notice that the electricity readings were taken monthly by staff but the gas consumptions were based upon the quarterly

Table 9.4 Initial gas and electricity consumptions for case study 9.2

Month	Gas (ft^3)	Electricity (kWh)	Month	Gas (ft^3)	Electricity (kWh)
September	–	5320	March	–	7000
October	–	6450	April	–	6375
November	233 364	6675	May	419 327	5815
December	–	6850	June	–	4140
January	–	7305	July	–	4320
February	587 058	7380	August	27 347	4370

readings from a gas meter registering in hundreds of cubic feet which have been adjusted accordingly in Table 9.4. Quarterly charges have been omitted to aid clarity. Clearly the initial audit will yield a limited analysis of energy consumption for this site. Table 9.5 shows the analysis.

Table 9.5 is set out and the figures calculated in a similar manner to Table 9.1.

COMMENTS ON THE INITIAL ANALYSIS OF CASE STUDY 9.2

The comments which can be made here are limited to the annual total cost of energy at £9344 and the rather obvious conclusion that although electricity is only 16% of the total energy consumption it represents 46% of the annual energy cost. It would be far more beneficial if the ways in which gas and electricity were consumed were broken down so that a more detailed analysis could be undertaken. This would, however, require the investment in the purchase (or hire) and installation of metering equipment.

9.7 Presentation of data – a more detailed audit

There now follows the second part of the case study (case study 9.3) in which metering equipment was installed so that the following services could be analysed:

Table 9.5 Analysis of the initial energy audit for case study 9.2

Fuel	Gas	Electricity	Totals
Annual consumption	1267 095 ft^3	72 000 kWh	
Consumption in common units of kWh	386 464	72 000	458 464
Annual cost £ sterling	£5024	£4320	£9344
Cost/kWh in pence	1.3	6.0	
Percentage of total cost	54%	46%	
Percentage of total consumption	84%	16%	

- space heating to radiator circuits and the air handling unit serving the warm air system;
- hot water supply to the restaurant, showers and wash hand basins in the toilets;
- electricity supplies to the lighting circuits, lifts, mechanical services plant and small power.

Appendix 5 contains a list of metering and monitoring equipment currently available on the market.

Case study 9.3

The observations which were taken during the course of the year are given in Tables 9.6 and 9.7 The treated floor area for catering was found to be 25 m^2. On investigation it was found that the hot water supply was generated from centralized instantaneous gas direct-fired heaters while the space heating system had its own dedicated plant consisting of two boilers in parallel operating on sequence control.

The radiator system accounts for the structural losses in the building and the air handling unit provides tempered air to the building via supply and return ductwork thus accounting for the heat losses resulting from the rate of air change.

Table 9.6 shows the record of observations of heat energy and electricity consumptions.

Table 9.6 Annual record of observations for case study 9.3

| Month | Heat energy consumption (GJ) | | | | | Electricity consumption (kWh) | | | |
	Rads	AHU	Catering	Showers	Basins	Lighting	Lifts	Power	Services plant
September	–	–	5	2	1	1800	620	1800	1100
October	10	7	6.5	3.5	2.2	2020	630	2500	1300
November	80	40	6.5	3.5	2.2	2200	625	2450	1400
December	90	45	6.5	3.5	2.2	2250	625	2475	1500
January	85	40	6.5	3.5	2.2	2850	630	2475	1350
February	100	50	6.5	3.5	2.2	2860	620	2500	1400
March	85	40	6.5	3.5	2.2	2500	625	2475	1400
April	70	35	6.5	3.5	2.2	2000	625	2450	1300
May	20	15	6.5	3.5	2.2	1700	640	2475	1000
June	–	–	5	2	1	820	620	1900	800
July	–	–	5	2	1	700	620	1800	1200
August	–	–	5	2	1	800	620	1700	1250
Totals	540	272	72	36	21.6	22 500	7500	27 000	15 000

Grand totals: 941.6 GJ, 72 000 kWh

Table 9.7 Gas meter readings for case study 9.3

	Gas meter readings (ft^3)	
Month	Heating	Hot water supply
September	–	9116
October	25 525	14 585
November	169 554	14 585
December	173 200	14 585
January	171 377	14 585
February	198 724	14 585
March	175 935	14 585
April	148 588	14 585
May	51 048	14 585
June	–	9116
July	–	9116
August	–	9116
Totals	1113 951	153 144

Grand total: 1 267 095 ft^3 (compares with Table 9.5)

Table 9.7 shows the record of observations of gas consumption for space heating and hot water supply.

SOLUTION TO CASE STUDY 9.3

The data collected from the additional metering equipment installed in the building are given in Tables 9.6 and 9.7. You will see that there are two further gas meters, each of which has independently recorded the gas consumption for the heating and hot water supply. There are five heat meters installed to independently record the heat energy consumption for the radiator system, the tempered air system via the air heater battery and the hot water supply to restaurant, showers and wash hand basins.

There are also subsidiary electricity meters for independently recording the electricity consumption for the lighting, lifts, power requirements and services plant. Clearly this will produce an energy audit with much more detailed analysis. Table 9.8 gives the analysis of the observations recorded in Table 9.6.

COMMENTS UPON THE ANALYSIS IN TABLE 9.8

The annual consumptions in the table are effectively the energies consumed at the points of use if one ignores distribution losses. The total energy used for heating and HWS from Table 9.6 is 941.6 GJ. The totals of energy used for heating and HWS in Table 9.8 are

Table 9.8 Common energy consumption totals from record of observations in Table 9.6.

Fuel	Gas heating			Gas HWS			Electricity (kWh)		
Service	Rad. system	AHU	Catering	Showers	Basins	Lighting	Lifts	Power	Services plant
Annual consump.	540 GJ	272 GJ	72 GJ	36 GJ	21.6 GJ	22 500	7500	27 000	15 000
Consumption (kWh)	150 000	75 556	20 000	10 000	6000	22 500	7500	27 000	15 000

Grand totals: Heating 225 556 kWh; HWS 36 000 kWh

225 556 kWh and 36 000 kWh respectively – a grand total of 261 556 kWh. This is equivalent to 941.6 GJ, thus agreeing with the earlier calculation.

SEASONAL EFFICIENCIES OF PLANT

If the energy consumption figures for gas in Table 9.7 are considered, the gross values of energy input of gas in Table 9.5 of 1267 095 ft^3 or 386 464 kWh can be split into that required for the space heating and that needed for the hot water supply. In Table 9.8 the net grand totals of heat energy used for space heating and hot water supply are given. In Table 9.7 the gross grand totals of gas consumed are given. The seasonal efficiencies can therefore be determined for the heating and the HWS from this data and these are shown in Table 9.9.

You will see that the total gas consumption in Table 9.9 agrees with the figure in Table 9.5. The so-called points of use consumptions in kWh are the energy consumption totals obtained from Table 9.8.

You can see from Table 9.9 that the seasonal efficiency for the hot water supply system is high. This is to be expected for direct fired instantaneous HWS plant. Manufacturers of instantaneous direct fired HWS heaters advertise test efficiencies of around 90%. The seasonal efficiency for the space heating system of radiators is typical, although the test efficiencies of modern conventional boiler plant can also reach figures of around 90% on full load.

Table 9.9 Seasonal efficiency of space heating and HWS systems for case study 9.3

Item	Space heating	Hot water supply
Gas consumption (ft^3)	1 113 951	153 144 (total = 1 267 095)
Conversion to kWh	339 445	46 666 (from Table 9.7)
Points of use (kWh)	225 556	36 000 (from Table 9.8)
Seasonal efficiency	66%	77%

Table 9.10 Performance indicators for the services identified in case study 9.3

Service	Net consumption (kWh)	Seasonal efficiency	Gross consumption (kWh)	Treated area (m²)	PI (kWh/m²)
Rad system	150 000	66%	227 273	2000	114
AHU	75 556	66%	114 479	2000	57
Catering	20 000	77%	25 974	25	(1039)
Showers	10 000	77%	12 987	2000	6.5
Basins	6000	77%	7792	2000	3.9
Lighting	22 500	–	22 500	2000	11.25
Lifts	7500	–	7500	2000	3.75
Power	27 000	–	27 000	2000	13.5
Services plant	15 000	–	15 000	2000	7.5

Total gross consumption 460 505 kWh

Total PI, excluding catering 217 kWh/m²

0.78 GJ/m²

DETERMINATION OF PERFORMANCE INDICATORS

A series of performance indicators can be generated from the data in Tables 9.8 and 9.9. The treated floor area for the building is given as 2000 m² and if the seasonal efficiencies from Table 9.9 are used as average efficiencies for each point of use for the heating and hot water supply systems, the performance indicators for the services can be determined as shown in Table 9.10.

COMMENTS ON THE ANALYSIS OF CASE STUDY 9.3 IN TABLE 9.10

The performance indicator for the showers is derived from the total treated floor area, inferring that the gymnasium and aerobics room are used by the occupants of the whole building, which may not be the case. Further investigation into the use of these areas would be helpful.

The performance indicator for the catering services is the consumption of natural gas; it does not include electrical power consumption.

The PI for catering compares with the energy consumption yardsticks given in Table 9.11 and taken from the Energy Efficiency Office's *Introduction to Energy Efficiency in Catering Establishments*. See Appendix 4.

You should now refer to Chapter 6 in which energy targets expressed in kWh/m² for building services are given from the Building Analysis files in the CIBSE monthly journals [2]. These can be compared with the results in Table 9.10.

Table 9.11 ECYs from the EEO

Restaurant with bar:	1100–1250 kWh/m^2
Fast food restaurant:	480–670 kWh/m^2
Pub restaurant:	2700–3500 kWh/m^2

The gross energy consumptions calculated at each point of use in Table 9.10 now allow the determination of the percentage of total site consumption of energy for each point of use.

The percentage of energy consumption for the points of use for heating and HWS can also be determined, and using the annual cost for these services, the cost of each service at the point of use can be calculated. Refer to Table 9.12.

A similar calculation can be done for the electrical services. This analysis is also set out in Table 9.12.

With the completion of Table 9.12 a fairly detailed picture of energy use on this site now emerges. The audit has provided the annual energy consumption and cost at each point of use as well as the performance indicators for each of the services in the building.

ACTION POINTS

It is left to you to prepare a summary and action points for improving the energy performance of the services in this building. These need to

Table 9.12 Cost analysis of data from Tables 9.5 and 9.10

Point of use	Percentage site consumption	Percentage total	Total costs	Percentage heating and HWS	Cost at point of use
Radiators	49.4			58.5	£2939
AHU	24.9			29.4	£1477
Catering	5.6			6.7	£337
Showers	2.8			3.3	£166
Basins	1.7			2.0	£100
Sub totals		83.4	£5024	100	£5024
				Percentage electrical consumption	
Lighting	4.9			31.3	£1352
Lifts	1.6			10.4	£449
Power	5.9			37.5	£1620
Services plant	3.3			20.8	£899
Sub totals	100	15.6	£4320	100	£4320
Totals		100	£9344		

be included with the tables of audit for the client. Have a look initially at the comments made for case study 9.1 Also refer to Appendices 2 and 3.

Clearly the resolution of the data in Tables 9.5–9.12 can be generated using a spreadsheet or database to take out the tiresome long-hand calculations.

QUALIFYING REMARKS

Qualifying remarks in the report should include:

- There will be margins of error in the sub-metering equipment used to determine the seasonal efficiencies for the space heating plant and the plant generating the hot water supply.
- The demand for electricity is assumed to be equal to the consumption which implies a conversion efficiency of 100%. This does not identify the operating efficiencies of pumps, fans, lifts, etc.
- It is now left to you to identify further qualifying remarks for this audit.

This completes the analysis for case study 9.3.

9.8 Further source material

Recourse can be made to the Energy Efficiency Office's *Fuel Efficiency Booklets* which are currently freely available [3].

9.9 Chapter closure

You should now have a clear idea of what an energy audit is and the courses of action that can be taken following an audit. The measurement of the consumption of energy on the site is set out and the discussion on primary energy tariffs should assist in identifying the factors for consideration when negotiating with the energy suppliers. You are now able to prepare and present an energy audit for the site and you are aware how the audit can be extended to the point of use by employing metering equipment.

Monitoring and targeting 10

a, b	regression coefficients
CUSUM	cumulative sum deviation
d	temperature rise due to indoor heat gains
DD	Degree Day(s)
HTHW	high temperature hot water heating
HWS	Hot water supply
LPG	low pressure gas
LTHW	low temperature hot water heating
N	number of days in period under review
n	number of observations
r	correlation coefficient
SDD	Standard Degree Day(s)
t_b	base temperature/control temperature
t_i	indoor design temperature in °C
t_m	average 24-hour mean daily outdoor temperature in °C
x	independent variable
y	dependent variable

10.1 Introduction

When a building owner has had an energy audit completed for the previous year it is very important to follow this up with an energy agenda to ensure that interest in the management of energy is focused and does not peter out. The audit, in addition to setting out the annual costs of energy consumptions on site, can promote a focus in two areas of energy conservation:

- Reduction in the difference between annual energy demand and annual energy consumption.
- Reduction in the demand for energy to its lowest practical value.

The tariffs for primary energy might be considered a third focus. However this issue should be addressed immediately following the audit as a matter of course if it has not been done before, and suggestions on factors to consider in negotiating a tariff are included in Chapter 9.

The two other areas of focus are discussed in Chapter 7. The purpose of this chapter is to consider how energy consumption can be monitored and the ways in which checks can be made to verify that energy saving measures are yielding the estimated benefits of lower energy consumption. The process of monitoring will also identify whether the consumption of energy is following the expected trend.

10.2 Monitoring procedures

Monitoring the consumption of energy should follow an initial energy audit which is based upon a previous year's fuel invoices. It will meet three main objectives:

- it will provide a more detailed annual audit;
- it will provide data for more detailed analysis at points of energy use to establish patterns and variations;
- it will provide data for a system of continuous performance monitoring.

These objectives will assist in having the site's energy consumption under continuous scrutiny.

The first step in monitoring is to prepare a block diagram. This identifies the locations of energy input, the plant which converts it for use, the media used, if any, for transportation to site locations and the services provided at the point of use. Figure 10.1 is a typical block diagram for site space heating, hot water supply, lighting and power.

The energy audit will have established the energy inputs. The block diagram should identify energy conversion plant and prime movers, transport media and distribution, and the services provision. In the process its preparation should give an insight into the extent and size of the systems which indirectly use fossil fuels on the site.

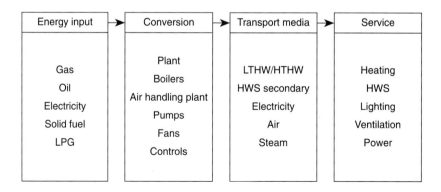

Figure 10.1 Block diagram for site space heating, HWS, lighting and power.

Having identified the points of energy input on the site an investment is usually required for monitoring and recording equipment to be located at the points of energy use, such as gas, oil, LPG and electricity. Most of this equipment can be hired for use. It includes multichannel data loggers, check meters, hours run meters, oil/water flow meters, gas meters, temperature sensors and a flue gas analyser for combustion efficiency tests. You should now refer to Appendix 5 for a more extensive list.

The frequency and accuracy of recording consumption data will depend upon the level of commitment. The reading and recording of the main meters and fuel deliveries must be a priority and should not be left to the fuel suppliers. Readings should be taken monthly, if not weekly, preferably on the same day and at the same time. A data record form should be devised for this purpose. The energy management record form shown in Figure 10.2 is reproduced from a CIBSE publication [1].

10.3 Monitoring equipment

Performance monitoring involves taking a number of pairs of readings for the purpose and plotting them on a graph to generate a thermal performance line which visualizes system performance. Two common pairs of readings used to measure the performance of space heating systems are:

- energy/fuel consumption and Degree Days;
- energy/fuel consumption and average mean daily outdoor temperature.

The correlation coefficient can have two functions:

- it establishes whether or not there is a degree of association between two variables;
- it validates the accuracy of the observations.

Clearly observations of fuel consumption and mean outdoor temperature for a building will have a strong degree of association. It is therefore the second function of the correlation coefficient which is needed to validate the accuracy of the observations taken. Table 10.1 is taken from a CIBSE publication [1] and lists minimum correlation coefficients for acceptable levels of association between two variables against the number of observations.

You will see that as the number of events increases so the value of the minimum correlation coefficient reduces.

If the calculated coefficient is below the minimum value given in Table 10.1, the observations are not valid, which is to say they are not sufficiently accurate. This means that more care must be taken in the process of recording fuel consumption and mean daily outdoor temperature.

10.4 Correlation and linear regression analysis

JAN	FEB	MAR	1st Q	APR		NOV	DEC	4th Q	YEAR			ELECTRICITY
										(1)	Month	
										(2)	Heating Degree Days	
										(3)	Lighting and General Power exc. Lifts (kWh)	ELECTRICITY
										(4)	Lifts (kWh)	
										(5)	Boiler House (kWh)	
										(6)	Total (3) + (4) + (5) (kWh)	
										(7)	Total (6) x 3.6 x 10^{-3} (GJ)	
										(8)	Maximum Demand (kVA or kW)	
										(9)	Maximum Demand Charge (£/kVA or £/kW)	
										(10)	Total Maximum Demand Charge (8) x (9) £	
										(11)	Unit Charge (£/kWh)	
										(12)	Total Unit Costs (6) x (11) (£)	
										(13)	Standing Charge(s) (£)	
										(14)	Total Electrical Costs (10) + (12) + (13) (£)	
										(15)	Quantity (litres or tonnes)	OIL OR COAL
										(16)	Calorific Value (MJ/litre or MJ/tonne)	
										(17)	Unit Cost (£/litre or £/tonne)	
										(18)	Energy input (15 x 16) x 10^{-3} (GJ)	
										(19)	Metered Output (GJ)	
										(20)	Plant Efficiency {(19) ÷ (18)} x 100 (%)	
										(21)	Total Cost (15) x (17) (£)	
										(22)	Quantity (therms)	GAS
										(23)	Unit Cost (£/therm)	
										(24)	Energy Input (22) x 0.1055 (GJ)	
										(25)	Metered Output (GJ)	
										(26)	Plant Efficiency {(25) ÷ (24)} x 100 (%)	
										(27)	Total Cost (22) x (23) (£)	
										(28)	Total Useable Energy (7) + (19) + (25) (GJ)	
										(29)	Total Site Energy (7) + (18) + (24) (GJ)	
										(30)	Total Cost (14) + (21) + (27) (£)	

Figure 10.2 Energy management record form. Reproduced with kind permission of CIBSE.

Table 10.1 Minimum correlation coefficients for acceptable levels of association; the correlation coefficients will have positive or negative values

Number of observations	Minimum correlation coefficient
10	± 0.767
15	± 0.641
20	± 0.561
25	± 0.506
30	± 0.464
35	± 0.425
40	± 0.402

Correlation is the degree of association between two unrelated quantities which are varying together. The correlation coefficient, r, is the measure of the association and varies from 0.99 downwards. A significant correlation is established if the coefficient is in excess of a given value which varies with the number of events or observations in the sample.

In research work any two variables can therefore be compared and the correlation coefficient calculated to see if there is an association between them. It is apparent that there should be a significant correlation between the pairs of observations suggested above. This will be tested in the following analysis. When they are plotted on a graph it is likely that there will be a scatter of points and hence some difficulty in placing the line of best fit. It is at this juncture that the adoption of another mathematical tool will be useful.

Case study 10.1

Consider the data in Table 10.2 relating to monthly energy consumption by a space heating plant and the monthly Degree Days for the locality.

SOLUTION

You will notice from Table 10.2 that in the summer months of July, August and September there is a constant load of 100 GJ.

Table 10.2 Monthly Degree Days and energy consumption in GJ for a consumer

Month	J	F	M	A	M	J	J	A	S	O	N	D	Total
DD	340	370	280	230	160	40	–	–	–	110	240	315	2085
Energy (in GJ)	330	380	310	240	170	150	100	100	100	180	340	270	2670

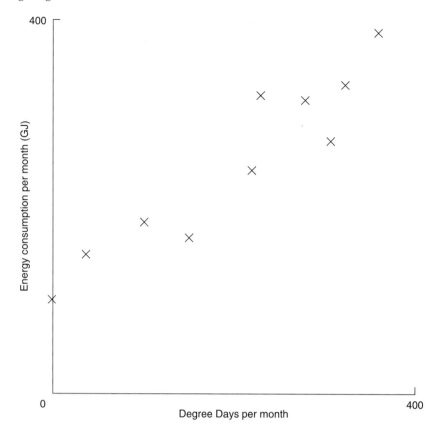

Figure 10.3 Typical scatter of observations.

It is clear that this load extends throughout the year and represents the requirement for hot water supply which is unrelated to the monthly SDD. This is known as the base load.

If these pairs of readings are plotted on a graph of Degree Days against energy consumption, a scatter of points will emerge as shown in Figure 10.3 and difficulty will be experienced in plotting the line of best fit.

The solution involves finding an equation which describes the correlation between the monthly Degree Days, x, and the monthly energy consumption, y. Degree days, x, is the independent variable, and that part of the energy consumption, y, which is weather related is therefore dependent upon the Degree Days and is known as the dependent variable. The equation of the regression line is thus y on x.

The technique of finding the equation is known as regression analysis. Linear regression, which applies here, implies a straight line association between x and y and the equation will therefore be of the form:

Table 10.3 Tabulated data based on information in Table 10.2

x	y	xy	x^2	y^2
340	330	112 200	115 600	108 900
370	380	140 600	136 900	144 400
280	310	86 800	78 400	96 100
230	240	55 200	52 900	57 600
160	170	27 200	25 600	28 900
40	150	6 000	1 600	22 500
–	100	–	–	10 000
–	100	–	–	10 000
–	100	–	–	10 000
110	180	19 800	12 100	32 400
240	340	81 600	57 600	115 600
315	270	85 050	99 225	72 900
$\Sigma x = 2085$	$\Sigma y = 2670$	$\Sigma xy = 614\ 450$	$\Sigma x^2 = 579\ 925$	$\Sigma y^2 = 709\ 300$

$$y = ax + b \tag{10.1}$$

where a and b are the regression coefficients. Values of a and b may be found from a pair of simultaneous equations:

$$\Sigma y = a\Sigma x + nb \tag{10.2}$$

$$\Sigma xy = a\Sigma x^2 + b\Sigma x \tag{10.3}$$

where n is the number of events or observations. The values of x and y are shown in Table 10.3.

The two simultaneous equations (10.2) and (10.3) now become:

$$2670 = 2085a + 12b$$

$$614\ 450 = 579\ 925a + 2085b.$$

Multiplying the first equation by 174:

$$464\ 580 = 362\ 790a + 2085b;$$

subtracting the third equation from the second:

$$149\ 870 = 217\ 135a$$

from which $a = 0.69$.

Substituting for a in the first equation:

$$b = 103.$$

The regression equation (10.1) is therefore:

$$y = 0.69x + 103.$$

Two arbitrary values can now be given to x within the range from zero to 400 and y evaluated. Let $x = 100$. Then $y = 172$. Let $x = 350$. Then $y = 345$. The two coordinates on the graph can then

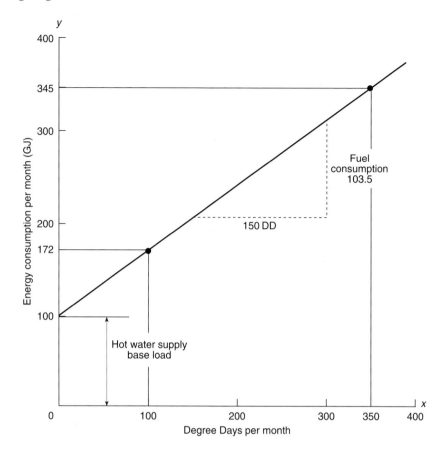

Figure 10.4   Graph of $y = 0.69x + 103$ for case study 10.1.

be connected by a straight line which will be the line of best fit in the scatter of points resulting from the observations. See Figure 10.4.

You will see that the regression coefficient b is evaluated as 103. In fact, of course, it should be 100 which is the energy consumption for the hot water supply. The error occurs in rounding the numbers in the simultaneous equations. As the consumption of hot water supply is consistent and continuous throughout the year it is called the base load which is weather unrelated. In practice the base load consumption of 100 GJ would not be exactly the same each month for the hot water supply. There would also be shut down periods for holidays which are not considered here.

It is now necessary to determine the correlation coefficient using the observations in Table 10.2. Clearly one would expect a close correlation between energy consumption and Degree

Days. The second function of the coefficient therefore applies here in validating the accuracy of the observations in Table 10.2. The value of the minimum correlation coefficient is dependent upon the number of observations. It is also dependent here on the association of energy consumption and Degree Days. The energy consumption for hot water supply is therefore unrelated. However, if it is consistent throughout the year, it is only necessary to omit the summer months of July, August and September when no heating was required. Table 10.3 must therefore be adjusted by omitting the three summer months, making the number of observations 9 instead of 12. This will affect the summations of y and y^2 in Table 10.3.

The formula for the calculation of the coefficient r is:

$$r = (\Sigma xy - n\bar{x}\bar{y})/\sqrt{((\Sigma x^2 - \bar{n}x^2)(\Sigma y^2 - n\bar{y}^2))} \qquad (10.4)$$

where $\bar{x}$ and $\bar{y}$ are the mean values and $n = 9$.

From Table 10.3, the summation over y will be $\Sigma y = 2670 - 300 = 2370$ and the summation over y^2 will be $\Sigma y^2 = 709\ 300 - 30\ 000 = 679\ 300$. Substituting these values and the other summations from Table 10.3 into equation (10.4)

$$r = 614\ 450 - (9 \times 231.66 \times 263.33)/\sqrt{(579\ 925 - 483\ 025)}$$
$$(679\ 300 - 624\ 100)$$
$$= 65\ 401/\sqrt{(96\ 900 \times 55\ 200)}$$
$$= 65\ 401/73\ 136$$
$$= +0.894.$$

From Table 10.1, which identifies minimum values for the correlation coefficient, 10 observations have a minimum value for r as 0.767. It is therefore apparent that the observations given in Table 10.2 are validated and therefore accurate.

In the determination of the correlation coefficient you may feel unhappy about including the hot water supply base load in the

Table 10.4 Tabulated data excluding the base load

x	y	xy	x^2	y^2
340	230	78 200	115 600	52 900
370	280	103 600	136 900	78 400
280	210	58 800	78 400	44 100
230	140	32 200	52 900	19 600
160	70	11 200	25 600	4900
40	50	2000	1600	2500
110	80	8800	12 100	6400
240	240	57 600	57 600	57 600
315	170	53 550	99 225	28 900
$\Sigma x = 2085$	$\Sigma y = 1470$	$\Sigma xy = 405\ 950$	$\Sigma x^2 = 579\ 925$	$\Sigma y^2 = 295\ 300$

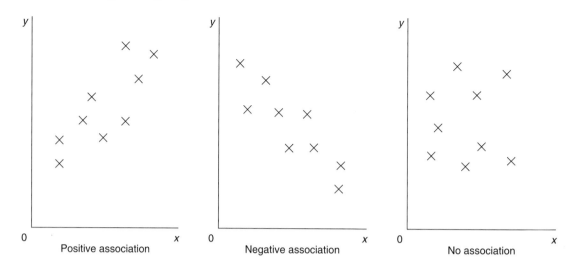

Figure 10.5 Scatter diagrams.

nine heating months, since it is unrelated to Degree Days. Table 10.4 discounts the hot water supply base load and you will see the subsequent calculation for r still agrees with its evaluation above.

Adopting the formula 10.4 for the correlation coefficient r:

$$r = 405\ 950 - (9 \times 231.67 \times 163.33)/$$
$$\sqrt{((579\ 925 - (9 \times 53\ 671))(295\ 300 - (9 \times 26\ 677)))}$$
$$= 65\ 400/\sqrt{(96\ 900 \times 5\ 200)}$$
$$= 65\ 400/73\ 136.$$
$$= +0.894.$$

Remember when two variables have an obvious association the second function of the correlation coefficient applies. If the observations were not recorded properly the correlation coefficient would draw attention to the fact by being below the acceptable minimum value given in Table 10.1. Care should therefore be taken in recording observations in a consistent manner and ensuring that, for example, accurate energy consumption readings are taken at the beginning or end of each calendar month to coincide with the monthly Degree Day totals.

The graphical effects of positive association, negative association and no association are shown on scatter diagrams in Figure 10.5.

SUMMARY OF PERFORMANCE ANALYSIS FOR CASE STUDY 10.1

- Slope of performance line is +0.69.

- Thermal energy intercept is 103 GJ (100 GJ).
- Correlation coefficient is +0.89 which implies a positive association. See Figure 10.5.

CONCLUSIONS

Regression analysis is a key mathematical tool for locating the line of best fit from a number of observations of two variables in a programme of monitoring energy consumption. Where the degree of association between two variables is not immediately apparent, determination of the correlation coefficient will provide the necessary evidence of association. It will also validate the accuracy of the observations. Regression analysis and correlation can be performed rapidly on a computer spreadsheet or database and on some pocket calculators.

In the drive for savings in energy the energy manager may want to investigate the possibility of a correlation between what at first sight might appear as two unrelated variables to establish if there is a degree of association between them. The following are potential examples:

- water consumption and mean outdoor temperature;
- power consumption and mean outdoor temperature;
- artificial lighting and levels of daylight.

10.5 Continuous performance monitoring using Degree Days

Case study 10.1 is an exercise in the historical performance of energy use for space heating and hot water supply, say for the previous year. The monthly performance of the space heating system over the nine months of the heating season can be obtained by dividing the GJ consumed by the Degree Days on a monthly basis as shown in Table 10.5.

The month of heaviest energy use was June, perhaps indicating that the plant was in need of servicing or it may have been cycling as a result of milder weather and hence low load. If the latter was the case, consideration could be given to introducing controls which inhibit boiler plant cycling under these conditions. November was also a poor month, but May at 0.44 GJ/DD provides the most efficient use of energy.

Table 10.5   GJ/DD for case study 10.1

Month	Jan	Feb	Mar	Apr	May	Jun	Oct	Nov	Dec	Total
DD	340	370	280	230	160	40	110	240	315	2085
GJ	230	280	210	140	70	50	80	240	170	1470
GJ/DD	0.96	0.76	0.75	0.61	0.44	1.25	0.73	1.0	0.54	0.705(av)

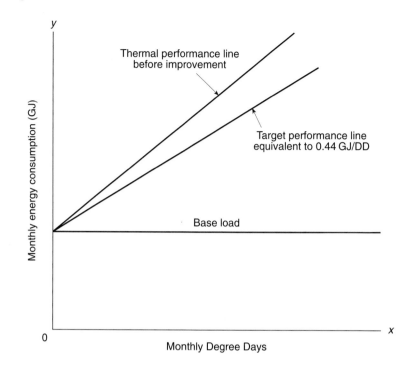

Figure 10.6 Target performance line.

The monthly variations in GJ/DD over the season's observations would give cause for concern and an investigation might be in order. The plant and system should be able to provide a monthly GJ/DD approaching the best historic value of 0.44. If this is achieved in the following year the slope of the performance line in Figure 10.4 would change.

For monitoring the continuous performance of the space heating system a target performance line could be drawn at a slope equivalent to 0.44 GJ/DD for the following year as shown in Figure 10.6.

It may be that an improvement of 10% can also be achieved with the base load energy consumption for hot water supply in which case the intercept at the y-axis would be at an average of 90 GJ/month instead of 100 GJ/month. The prevailing performance line can then be added as the current year proceeds and compared with the target performance line.

The drawback here is that it is not until well into the prevailing year that any discrepancy is identified although the prevailing thermal performance line can be projected after the first two observations are entered and then adjusted as the prevailing year proceeds and more observations are available.

The correlation between fuel or energy consumption and Degree Days has been tested from the historical observations given in case study 10.1 and a close association confirmed for these two variables. It can also be shown that a close association should exist between energy consumption and the 24-hour mean value of outdoor temperature, preferably taken locally to the site. Mean outdoor air temperature can be measured daily from a maximum/minimum mercury in glass thermometer or its electronic counterpart in which case it can be automatically recorded on one of the output channels on a data logger and the weekly or monthly mean outdoor air temperature evaluated.

Alternatively the mean outdoor air temperature can be determined approximately from the weekly or monthly Degree Day totals for the locality from:

$$t_m = t_b - DD/N \qquad (10.5)$$

where

t_m = average 24-hour mean daily outdoor temperature (°C);
t_b = base temperature (taken as 15.5°C for SDD) (°C);
N = number of days in period under review;
DD = number of Degree Days to base d in period under review.

Monthly SDD totals are currently freely available from the Energy Efficiency Office's magazine [2].

The following case study analyses the thermal performance of a space heating plant with a hot water supply base load using average 24-hour mean daily outdoor temperature determined from weekly Standard Degree Days. It will be shown that by adopting outdoor temperature a more detailed analysis of system performance can be made.

Case study 10.2

Consider for analysis the tabulated observations in Table 10.6 of SDD/week and energy consumption in MJ/week which includes a base load of 100 MJ/week attributed to the hot water supply.

Table 10.6 Observations for analysis, case study 10.2

Week no.	1	2	3	4	5	6	7	8	9	10	11	12
SDD/week	79	91	31.5	70	21	61	3.5	55.3	13.3	47	24.5	17.5
MJ/week	475	525	290	360	150	312	195	245	110	335	300	240

Table 10.7 SDD/week converted to average 24-hour mean outdoor temperature

MJ/week	475	525	290	360	150	312	195	245	110	335	300	240
t_m	4.2	2.5	11	5.5	12.5	6.8	15	7.6	13.6	8.8	12	13

SOLUTION

Adopting the formula (10.5) for the average 24-hour mean daily outdoor temperature t_m:

$$t_m = t_b - DD/N$$

the average mean daily temperature for each week can be determined and this is shown in Table 10.7 together with the associated weekly energy consumption observations.

Taking the independent variable x as the average 24-hour mean daily temperature and the dependent variable y as the weekly energy consumption in MJ, Table 10.8 can now be generated.

Adopting the simultaneous equations (10.2) and (10.3) for linear regression and substituting for the twelve observations:

$$3567 = 112.5a + 12b$$

$$28\ 785 = 1236a + 112.5b.$$

Multiplying the first equation by 9.375:

$$33\ 441 = 1055a + 112.5b.$$

Subtracting the second equation we have:

$$4656 = -181a$$

Table 10.8 Tabulated data based on observations in Table 10.7

x	y	xy	x^2	y^2
4.2	475	1995	17.64	225 625
2.5	525	1313	6.25	275 625
11	290	3190	121	84 100
5.5	360	1980	30.25	129 600
12.5	150	1875	156.25	22 500
6.8	312	2122	46.24	97 344
15	195	2925	225	38 025
7.6	245	1862	57.76	60 025
13.6	110	1496	185	12 100
8.8	335	2948	77.44	112 225
12	330	3960	144	108 900
13	240	3120	169	57 600
Σx 112.5	Σy 3567	Σxy 28 785	Σx^2 1236	Σy^2 1 223 669

from which $a = -25.7$.

Substituting for a in the first equation:

$$3567 = -2891 + 12b$$

from which $b = 538$.

The regression equation (10.1) is therefore:

$$y = -25.7x + 538.$$

If x is now given two values, say 5 and 15, y is calculated from this regression equation as 409.5 and 152.5, respectively. These values can now be plotted on a graph of the average 24-hour mean daily outdoor temperature x against weekly energy consumption y and the best fit thermal performance line drawn as shown in Figure 10.7.

You may also now like to plot the 12 observations on the graph to see how the performance line fits into the scatter of points. The results of the system performance analysis are:

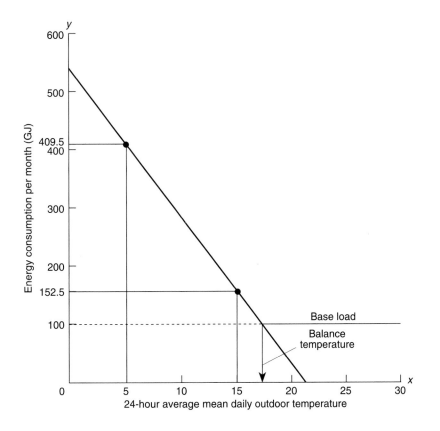

Figure 10.7 Thermal performance line for case study 10.2.

slope of thermal performance line is −25.7;
temperature intercept is 21°C;
energy intercept is 540 MJ/week;
balance temperature is 17°C.

- If the negative numerical value of the slope increases, the performance line becomes steeper and this signifies an improvement in
  system performance.
- If the negative numerical value of the slope decreases, the slope of
  the thermal performance line becomes flatter and this signifies
  poorer system performance. See Figure 10.7.
- The energy intercept indicates the energy consumption when the
  average 24-hour mean daily outdoor temperature is calculated as
  zero.
- The balance temperature is the outdoor temperature at which no
  heating is required; it is also the base temperature t_b for the building
  where

$$t_b = t_i - d$$

and d is the temperature rise due to indoor heat gains. Thus base
temperature is the indoor temperature above which no heating is
required. See Chapters 2 and 3. The base temperature is also called
the control temperature.

If the balance temperature identified from the intersection of the
thermal performance line with the base load is in excess of the base
temperature for the building, the calculated indoor heat gains may be
in error or the space heating system is not functioning efficiently. The
latter will certainly be the case if the balance temperature is greater than
the design indoor temperature. This may be due to:

- poor system performance resulting from inadequate space temperature controls, incorrect adjustment of space heating controls
  or malfunctioning of the space heating controls;
- low boiler efficiency due to poor combustion conditions;
- poor control of boiler plant causing unnecessary boiler on/off
  cycling.

The correlation coefficient can be determined from the summations in
Table 10.8.

It is not necessary here to extrapolate the base load from the weekly
energy figures as it was in case study 10.1. The reason for this is that
the 12-week period over which the observations were taken did not
include summer time when the space heating is shut down. In case study
10.1 a calendar year's observations were taken which included the
summer shut down period for the space heating system.

Therefore from the correlation coefficient equation (10.4) and the
summations in Table 10.8,

$$r = 28\,785 - 12\,(112.5/12)\,(3567/12)/\surd(1236 - 12(112.5/12)^2)$$
$$\times\,(1\,223\,669 - 12(3567/12)^2)$$

from which

$$r = -4656/\surd(181.3 \times 163\,378)$$
$$= -0.855.$$

The negative sign implies a negative association; see Figures 10.5 and 10.7. From Table 10.1 the minimum correlation coefficient for there to be an association between the variables of the average 24-hour mean daily outdoor temperature and weekly energy consumption is about ±0.7. If the coefficient is calculated to be outside the minimum value the observations on which it is based are inaccurate and therefore the thermal performance line is invalid. More care would be needed in recording the observations and checks made on the measuring and recording equipment.

10.7 Correcting fuel/energy consumption to a common time base

Data from observations of energy consumption and Degree Days or mean daily outdoor temperature should always give a consistent indication of energy use.

Sometimes occupancy and hence hours of use vary from one period to the next.

It is important to check, for example, that an office space heating plant operates for regular periods each day of the week. Optimum start/stop time control of plant will have little effect except perhaps on Mondays after a weekend shut down.

Consider observations of fuel consumption and mean daily temperature taken for a sports hall in case study 10.3.

Case study 10.3

The school sports hall has a heating system independent of the rest of the campus.

It is heated by high temperature gas fired radiant tube and the observations are recorded in Table 10.9.

Table 10.9 Observations for case study 10.3

Week no.	1	2	3	4	5	6	7
Period (hours)	35	30	30	40	30	45	30
Energy (GJ)	23.3	25	30	37.3	27	33	15
t_m (°C)	8	8.5	6	7.4	6.6	9.2	10

Table 10.10 Observations corrected to a common time period

Week no.	1	2	3	4	5	6	7
Energy (GJ)	20	25	30	28	27	22	15
t_m (°C)	8	8.5	6	7.4	6.6	9.2	10

SOLUTION

As the period of 30 hours is more regular the remaining observations can be corrected to the base of 30 hours:

Week 1: (23.3 GJ/35 hrs) × 30 = 20 GJ
Week 4: (37.3 GJ/40 hrs) × 30 = 28 GJ
Week 6: (33 GJ/45 hrs) × 30 = 22 GJ.

The revised observation record is shown in Table 10.10.
 You may be able to deduce by inspecting the observations that there is a close association between the independent variables in Table 10.10. The scatter on a plot of t_m against energy consump-

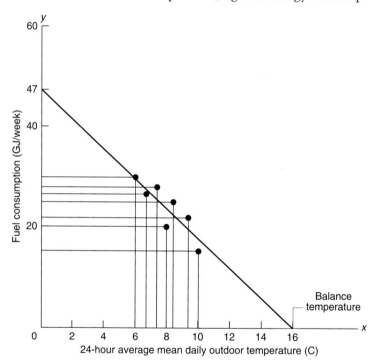

Figure 10.8 The thermal performance line for case study 10.3 plotted by trial and error.

tion should not be wide and therefore the line of best fit could be located manually without the need to undertake a linear regression analysis. Plotting the observations on graph paper to suitable scales will prove the matter. This is done in Figure 10.8.

The thermal performance line intersects the average 24-hour mean daily outdoor temperature axis and zero fuel consumption at approximately 16°C. This temperature should be the balance temperature for the sports hall at which no heating is required. The thermal energy intercept when the average 24-hour mean daily outdoor temperature is zero is 47 GJ.

On the other hand you may prefer to undertake the regression analysis and the determination of the correlation coefficient which would validate the observations. Clearly if you have this on a database or spreadsheet there is no argument in favour of plotting the line by trial and error. The linear regression equation from a solution by analysis is:

$$y = -3.25x + 50.$$

If the independent variable x is now given the values of, say, 2°C and 14°C, the values of the dependent variable y will be 43.5

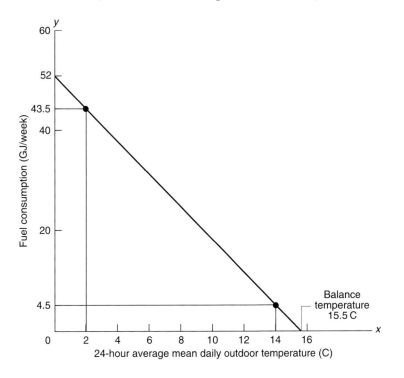

Figure 10.9 The thermal performance line for case study 10.3 plotted from the linear regression equation.

GJ and 4.5 GJ, respectively. These values are then plotted on a graph of x, the average 24-hour mean daily outdoor temperature, and y, the fuel consumption in GJ, as shown in Figure 10.9, from which the thermal performance line can now be drawn and the following data is obtained:

the slope of the thermal performance line is -3.25 GJ/K;
the temperature intercept (balance temperature) is $15.5°C$;
the thermal energy intercept at an average 24-hour mean
    daily outdoor temperature of zero degrees Celsius is 52 GJ;
the correlation coefficient is calculated as -0.916.

SUMMARY TO CASE STUDY 10.3

You should now undertake the determination of the regression equation and the correlation coefficient to confirm the solutions offered here. Clearly more information is obtained by undertaking a full analysis. The balance temperature is more accurately determined along with the slope of the thermal performance line. The energy use on a day of severe weather is predicted and the correlation coefficient is obtained. The minimum value of the coefficient for 10 observations from Table 10.1 is ± 0.767. Seven observations were obtained in the study, giving a coefficient of -0.916 which shows a close association between the two independent variables.

If the minimum value for the correlation coefficient for the number of observations as shown in Table 10.1 is not achieved, the energy survey is void and greater care over measurement and collection of data is required.

10.8 Performance monitoring using cumulative sum deviation

It is sometimes important to be able to demonstrate to the client that an energy saving measure for space heating is actually saving energy following its provision. This is particularly true on a site which consumes energy for other purposes in addition to space heating. Cumulative sum deviation achieves this objective and can even be used to identify the monthly and cumulative savings in the annual fuel account.

If CUSUM is adopted on a continuous basis it will show whether or not financial benefit resulting from the energy saving provision continues. CUSUM can also be adopted where energy is used for manufacturing products and in manufacturing processes. Case study 10.4 investigates the effects of introducing an energy saving measure on a space heating system by adopting CUSUM performance monitoring.

Table 10.11 Monthly Degree Day, energy and regression data for case study 10.4 one year before energy saving provision

Month	x	y	xy	x^2	y^2
September	51	10.0	510	2601	100
October	130	21.5	2795	16 900	462
November	137	18.5	2535	18 769	342
December	257	24.0	6168	66 049	576
January	309	39.5	12 206	95 481	1560
February	223	29.0	6467	49 729	841
March	277	32.0	8864	76 729	1024
April	157	23.0	3611	24 649	529
May	101	17.6	1778	10 201	310
June	–	5.0	–	–	25
July	–	5.0	–	–	25
August	–	5.0	–	–	25
	Σx 1642	Σy 230	Σxy 44 934	Σx^2 361 108	Σy^2 5820

Case study 10.4

The first two columns in Table 10.11 list the monthly Degree Days x and energy consumptions y in GJ in the year before the energy saving provision is put in place for an office located in the Thames Valley. The table is then extended in order to undertake a linear regression analysis for that year.

SOLUTION

Adopting the regression equations (10.2) and (10.3) and substituting the summations for twelve observations where $n = 12$:

$$230 = 1642a + 12b$$

$$44\ 934 = 361\ 108a + 1642b.$$

Multiplying the first equation by 136.8 we have:

$$31\ 464 = 224\ 626a + 1642b.$$

Subtracting this from the second equation we have

$$13\ 472 = 136\ 482a$$

from which $a = 0.1$.

Substituting for the regression coefficient a in the first equation we have

$$230 = 162 + 12b$$

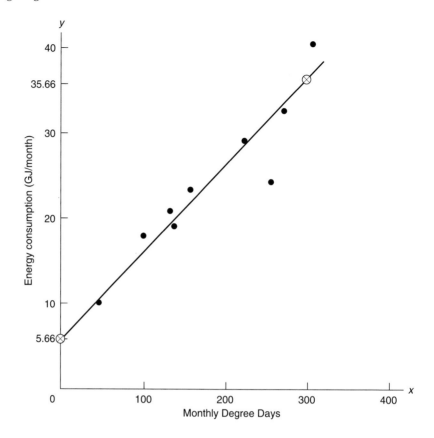

Figure 10.10 Case study 10.4: the thermal performance line in the year before energy saving provision

from which $b = 5.66$.

Thus the regression equation is $y = 0.1x + 5.66$.

The performance line of best fit can now be drawn for the year before the energy saving provision although it plays no part in the calculation of CUSUM. This is shown in Figure 10.10, for when $x = 0$, then $y = 5.66$ and when $x = 300$, $y = 35.66$.

Before proceeding further it is advisable to calculate the correlation coefficient from the weather related data in Table 10.11 to ensure that the observations are sufficiently accurate. There are nine observations of weather related data and the summations are listed in Table 10.12.

Table 10.12 Summations of weather related data for case study 10.4

Months	x	y	xy	x^2	y^2
September to May	Σx 1642	Σy 215	Σxy 44 934	Σx^2 361 108	Σy^2 5745

You will notice in Table 10.12 that only summations for y and y^2 have altered from Table 10.11 on account of the summer months when the energy consumptions are not weather related.

The correlation coefficient r for the weather related observations in the first year can now be determined from equation (10.4) and substituting we have:

$$r = (44\ 934 - 9(1642/9)(215/9))/\sqrt{(361\ 108 - 9(1642/9)^2)}$$
$$(5745 - 9(215/9)^2)$$
$$= (44\ 934 - 39\ 226)/\sqrt{(361\ 108 - 299\ 559)(5745 - 5136)}$$
$$= 5708/\sqrt{(61\ 549 \times 609)}$$
$$= 5708/6122$$

from which the correlation coefficient $r = +0.93$.

The minimum value for the coefficient for nine observations is, from Table 10.1, about ± 0.8. Thus the observations are validated and we can proceed.

Table 10.13 CUSUM for case study 10.4

Month	DD X	Actual energy y	Predicted energy	Difference	CUSUM
September	51	10	10.76	−0.76	−0.76
October	130	21.5	18.66	2.84	2.08
November	137	18.5	19.36	−0.86	1.22
December	257	24	31.36	−7.36	−6.14
January	309	39.5	36.56	2.94	−3.20
February	223	29	27.96	1.04	−2.16
March	277	32	33.36	−1.36	−3.52
April	157	23	21.36	1.64	−1.88
May	101	17.6	15.76	1.84	−0.04
June	–	5	5.66	−0.66	−0.70
July	–	5	5.66	−0.66	−1.36
August	–	5	5.66	−0.66	−2.02
September	55	8.7	11.16	−2.46	−4.48
October	138	13	19.46	−6.46	−10.94
November	241	17	29.76	−12.76	−23.7
December	299	22	35.56	−13.56	−37.26
January	337	28	39.36	−11.36	−48.62
February	308	28.3	36.46	−8.16	−56.78
March	270	21.5	32.66	−11.16	−67.94
April	199	18	25.56	−7.56	−75.50
May	114	11.2	17.06	−5.86	−81.36
June	–	5	5.66	−0.66	−82.02
July	–	5	5.66	−0.66	−82.68
August	–	5	5.66	−0.66	−83.34

The calculation of the cumulative sum deviation for the office can now be undertaken. This is listed in Table 10.13, which includes the monthly Degree Days and energy consumptions for both the year before and the year following the introduction of the energy saving provision.

The column of predicted energy consumption is calculated using the regression equation determined from the observations made in the year prior to the introduction of the energy saving measure. The equation is $y = 0.1x + 5.66$.

The column headed 'difference' identifies the difference between actual energy consumption y and predicted energy consumption. A minus sign indicates that the predicted energy consumption exceeds the actual monthly consumption of energy.

The last column is the cumulative effect (CUSUM) of the 'difference' column over the two consecutive years.

The CUSUM values in the last column of Table 10.13 are now plotted against the months of the two-year period and the result is shown in Figure 10.11.

ANALYSING THE PLOT OF CUSUM IN FIGURE 10.11, CASE STUDY 10.4

- The graph is based upon the regression equation calculated from the observations prior to the energy saving provision.
- You will see that the plot hovers around the base line during the year preceding the implementation of the energy saving provision. This is to be expected since the regression equation used for all the predictions was calculated from that year's observations.
- The deviation begins in the September of the second year and continues until the following May at the end of the heating season. The deviation reverts at this point back to a second base line. This occurs because the office base load which accounts for the consumption of hot water is weather unrelated and therefore not influenced by Degree Days.
- The deviation which occurs between the September of the second year and the following May identifies the cumulative savings in the monthly energy consumption, the total being in the region of 83 GJ for that period. This can easily be costed progressively each month starting in the September of the second year.
- If the CUSUM calculation is continued into the year after and beyond, a constant watch can be taken to ensure that the savings continue. If the slope of the line of deviation in Figure 10.11 changes during the following heating season, it shows that either further energy savings have been introduced (steeper slope) or that the initial energy saving provision is lapsing (flatter slope) or that something else has caused an increase in energy consumption.

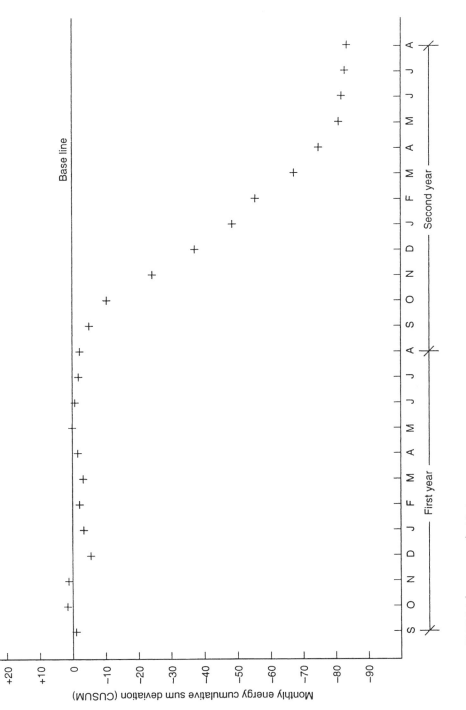

Figure 10.11 CUSUM for case study 10.4.

- CUSUM cannot divulge what the energy savings or increased energy consumption can be attributed to. In fact in this case study which relates to an office block in the Thames Valley, temperature controls were upgraded and commissioned in the August at the end of the first year of observations. As long as the temperature controls are maintained in accordance with operating procedures, and preventative maintenance to the building and to the services is ongoing, the savings in energy should continue. This will be confirmed by the CUSUM deviation slope.

10.9 Chapter closure   You now have the underpinning knowledge to undertake a variety of monitoring and targeting procedures. You can analyse the results of monitoring energy consumption and prepare documentation to support evidence of the effects of energy conservation measures, and documentation to confirm savings in the cost of energy consumed for space heating and hot water supply. A list of typical energy conservation measures is included in Appendix 2.

Appendix 1
Standard heating
Degree Day data

Degree Days can show for a given heating season how far outdoor temperatures are on average below the base temperature (known also as the control temperature or the balance temperature) which is taken as 15.5°C. When the outdoor temperature is above the base temperature it should not be necessary to heat the building. This is because of the effect of indoor heat gains, d, caused from heat generated by the occupants of the building and also that from equipment and lighting. The higher the annual number of SDDs the colder is the heating season.

SDDs are published by the Department of the Environment currently in the *Energy Management Journal* for every month of the year. The annual nine-month SDDs for the 20-year period to May 1979 are given in Table 1.5, Chapter 1, for 17 locations in the UK. The SDDs given in this appendix are, like those listed in Table 1.5, taken from 1 September to 31 May but from a later 20-year period.

Many organizations do not start up space heating plant now until 1 October, thus reducing the heating season from 273 days to 243 days. Account should be taken of this when using the SDD data. For this reason there is included here the SDD for the month of September averaged over 20 years to 1993 so that it can be deducted from the annual nine-month totals if required.

Area location	Location	20-year average to May '93 annual SDD	Sept '94 to May '95 annual SDD	20-year average to '93 SDD Sept only
Thames Valley	Heathrow	1961	1650	55
South Eastern	Gatwick	2171	1829	78
Southern	Bournemouth	2100	1801	79
South West	Plymouth	1775	1541	57
Severn Valley	Bristol	1868	1577	53
Midlands	Birmingham	2297	2035	88
West Pennines	Manchester	2197	2003	83
North Western	Carlisle	2343	2175	104
Borders	Boulmer	2376	2229	107
North Eastern	Leeming	2317	2122	90
East Pennines	Finningly	2265	2067	85
East Anglia	Honington	2252	2030	78
West Scotland	Glasgow	2402	2237	114
East Scotland	Leuchars	2463	2346	115
N. E. Scotland	Aberdeen	2544	2413	128
N. W. Scotland	Stornoway	2396	2399	135
Wales	Aberport	2089	1860	78
N. Ireland	Belfast	2290	2118	105

Notes:
(i) The location area of N. W. Scotland has been included here. It is not listed in Table 1.5, Chapter 1.
(ii)The annual SDDs recorded from September 1994 to May 1995 are generally lower than the 20-year average to 1993, denoting a warmer climate for that year. This may indicate a trend or it may be a blip.
(iii) It would be of interest to undertake a comparison between the 20-year average SDD to May 1979 listed in Table 1.5 with the 20-year average SDD to May 1993 listed in this appendix, in order to establish a trend. This comparison can be illustrated graphically by plotting on the same graph the location areas against the annual SDD for each location for the two periods.

Appendix 2
Energy conservation measures

There are three matters which must be investigated prior to a consideration of energy conservation measures.

- There is in place an active preventive maintenance programme for the services within the building. See Appendix 3.
- There is in place an active preventive maintenance programme for the building envelope.
- The occupants of the building are satisfied with the level of comfort provided by the services in the building.

- Roof insulation equivalent to 150 mm of glass fibre having a thermal conductivity of 0.035 W/m K.
- Wall insulation by dry lining, cavity fill or external cladding to a thermal transmittance coefficient of 0.4 W/m^2 K or less.
- Single glazing changed to double glazing having at least a 12 mm cavity and the facility for trickle ventilation in the window frame if the building is not air conditioned.
- Sealing around doors and windows.
- Air locks provided at entrance doors.
- Automatic closure of fire safety doors on corridors off the lifts and stairwells.
- Ventilation of the lift shaft is from outdoors to outdoors and not affected by strong wind.
- Toilet extract systems do not operate continuously but in response to operation of the artificial lighting system in the toilet or to door opening.

**Building services –
space heating**

- Check the time scheduling with the occupation times and the thermal response factor for the building. Optimum start–stop controls may be appropriate.
- If parts of the building have different occupation times consider local time scheduling via two or three port zone valves.
- Check thermostat settings on boiler plant and zones.
- The minimum temperature control for radiator systems is weather compensated on the boiler plant or weather compensated via a constant volume variable temperature control on a three port valve.
- There should be local temperature control available via thermostatic radiator valves or two port zone valves.
- Recent developments in boiler design have substantially increased the efficiency of heat conversion and at the same time reduced carbon dioxide and nitrous oxide emissions. Replacement should therefore be seriously considered if boiler plant is more than 10 years old.
- If there is only one heating boiler ask why.
- If there is more than one boiler, are sequence controls provided?
- Are time delays fitted to prevent the boiler plant from starting on a sudden temporary demand?
- Is the space heating boiler plant independent of the generation of hot water supply?
- If there is more than one building on the site is there a central boiler plant room? If so consider the provision of local plant when renewal is on the agenda.
- Check the thermal insulation on distribution pipes.
- If windows are opened during cold weather find out why.

**Building services –
hot water supply**

- The generation of hot water supply should be independent of the space heating plant.
- Heat loss is sustained if the hot water is stored in a vessel prior to consumption – investigate the use of direct fired instantaneous heaters.
- Check that hot water is only provided during the occupied period.
- Check the thermal insulation of storage vessels and distribution pipes.
- Check the storage or operating temperature which should be a minimum of 60°C.
- Ensure that secondary circulation is taken to a point local to the draw-off point.
- Consider the use of spray taps on basins.

- Provide local heating and ventilation to the kitchen area.
- Ensure that local temperature controls are provided and used by the catering staff.
- Ensure that local ventilation controls are provided and used by the catering staff.
- Consider providing local heating for domestic hot water.
- Install separate metering for the kitchen equipment.
- Ensure that kitchen equipment is modern and energy efficient.

Catering services

- Check lighting levels in all areas with the recommended illuminance.
- Check the window cleaning schedule.
- Ensure that the luminaires are regularly cleaned and lamps replaced before they fail.
- Use energy efficient fluorescent tubes with electronic fittings.
- Ensure that GLS tungsten lamps are replaced if possible.
- Educate the occupants to switch the lights off.
- Install automatic lighting controls.

Artificial lighting

- Ensure that equipment with standby mode does not have this facility disabled.
- Ensure that staff switch off equipment not in use.
- Ensure that security staff switch off equipment not required when the building is unoccupied.
- Identify equipment which is not energy efficient.

Small power equipment

- Check the time scheduling with the occupation times.
- Check the air flow rates at the supply and extract grilles and of the fans in the air handling unit.
- Check the temperature and, where appropriate, the humidity control settings.
- If the system does not provide for space heating, must it operate at all times when the building is occupied?
- If the system operates on full fresh air only, consider the installation of a recuperator in the extract/supply air duct if this is not fitted.
- If there are substantial heat gains in the building is the sensible and latent heat being taken out from the extract air to heat the fresh air supply?

Mechanical ventilation

Appendix 3
Preventive
maintenance measures

Preventive maintenance, whilst not strictly part of an energy manager's remit, does play an essential role in controlling the consumption of electricity and fossil fuel on a site. Poor maintenance and reliance upon corrective maintenance will lead to increases in energy use. The energy manager is well advised to ensure that a programme of preventive maintenance is in place and strictly observed to the point where the customer builds into the programme his own checks and balances to verify the performance of the maintenance staff or contractors. This not only includes the presentation of signed certificates for each item of plant serviced, for example, but also witnesses the maintenance work being undertaken even if it is only on a spot check basis.

Another essential element associated with the proper operation of the mechanical and electrical services within the building is the initial commissioning of the systems. The energy manager would be wise to ensure that the commissioning process had been properly undertaken and recorded for each of the services in the building.

The following list of preventive maintenance measures cannot be exhaustive and does not suggest frequency, but hopefully gives an understanding of the breadth and depth of commitment required and all too frequently forgotten.

Space heating

- Plant room: should be clean and tidy and free from clutter. Check the fresh air intake.
- Boiler plant: cleaning and adjustment to fuel burner, combustion test using a flue gas analyser identifying products of combustion and quantities. Air fuel ratio check. Thermostat settings, sequence controls. Cleaning of boiler fire tubes, check on smokepipe and flueways. Thermal efficiency test. Check for leakage of flue gas and water.

- System: flushing out and use of inhibitor, feed and expansion tank, ball valve, open vent, overflow, cold feed, cover in place, thermal insulation/pressurization unit and feed pump.
- Controls and control valves, TRVs: check actuator, operation and settings and time scheduling.
- Pumps and fans: check for noise, speed, power supply, pressure developed, belt tension and wear, leakage, lubrication.
- Distribution pipes: thermal insulation, valves, drain points, air eliminators.
- Space heating appliances: clean heat exchangers and filters in fan convectors/fan coil units/unit heaters/natural draught convectors. Check the fans as for pumps and fans.
- Radiators: clean out convection channels, ensure that they are painted with non-reflective paint, clean reflective panels behind the radiators, check the radiator valves for leakage and operation.

- Calorifier/indirect cylinder: check the heat exchanger for corrosion/scale, check the thermal insulation.
- Secondary pump: check noise, speed, power supply, pressure developed, leakage, corrosion and scale.
- Storage tanks: check contents, ball valve, open vent, overflow, cold feed, cover in place, insulation.
- Secondary flow and return: flushing out to remove scale, thermal insulation, valves and stopcocks, drain points.

Central storage hot water supply

As for boilers and secondary pumps and secondary flow and return. In addition, since the heater is using raw water, the heat exchanger will require regular inspection. These heaters are frequently fed from the rising main and the following equipment should be subject to checking: strainer, pressure limiting valve, check valve, expansion valve, temperature/pressure relief valve, expansion vessel.

Direct fired hot water supply heaters

Due to their design, secondary water treatment and regular descaling of the plates may be necessary. An inhibitor should be considered for the primary water.

Plate heat exchangers for hot water supply

- Plant room: should be clean and free from clutter.
- Fans as for space heating pumps.
- Filters: cleaned/replaced.
- Fresh air intake: debris cleared away.
- Air handling unit: check air heater batteries and cooling coils for leaks and dirt build up, recuperators for cleanliness, volume

Mechanical ventilation

control dampers and linkages, and casing for leaks.

- Controls and controllers: check actuators, operation and settings, thermostat, humidistat settings, time scheduling.
- Distribution ductwork: check for cleanliness, check fire dampers, volume control dampers, thermal insulation.
- Supply and extract grilles/diffusers: check directional louver positions, flow rates, air temperatures.

Appendix 4
Energy Efficiency Office's series of booklets on *Introduction to Energy Efficiency*

The Energy Efficiency Office of the Department of the Environment has published, through BRECSU, the Building Research Energy Conservation Support Unit, a series of booklets to aid energy managers responsible for one or more of 13 different types of buildings. These booklets are currently freely available and are listed below.

Introduction to Energy Efficiency in Catering Establishments
Introduction to Energy Efficiency in Entertainment Buildings
Introduction to Energy Efficiency in Factories and Warehouses
Introduction to Energy Efficiency in Further and Higher Education
Introduction to Energy Efficiency in Health Care
Introduction to Energy Efficiency in Hotels
Introduction to Energy Efficiency in Museums, Art Galleries, Libraries and Churches
Introduction to Energy Efficiency in Offices
Introduction to Energy Efficiency in Post Offices, Building Societies, Banks and Agencies
Introduction to Energy Efficiency in Prisons, Emergency Buildings and Courts
Introduction to Energy Efficiency in Shops and Stores
Introduction to Energy Efficiency in Schools
Introduction to Energy Efficiency in Sports and Recreation Centres

The booklets each include information on energy management, the action plan, measures to achieve energy savings, energy use, energy consumption yardsticks and carbon dioxide yardsticks.

Appendix 5
Monitoring equipment

There is a wealth of equipment available for hire or purchase which can monitor and record. The Building Services Research and Information Association (BSRIA) will hire out equipment for monitoring and recording purposes. Some of this equipment is listed here.

- Meters: digital electricity meters, digital gas meters, digital heat meters.
- Heat meter simulation: heat energy consumption in space heating subcircuits can be determined in kWh or GJ by employing a portable ultrasonic flow meter and electronic thermometers clamped on to the circuit flow and return pipes and connected to a data logger.
- Air flow measurement: electronic vane anemometers, thermal vane anemometers, anemometer hoods (for measuring air flow from grilles), pitostatic tubes.
- Pressure measurement: micromanometers, Bourdon gauges, differential pressure gauge test sets, U-tube differential pressure test sets, static pressure transducers.
- Duct leakage tests: portable test sets.
- Temperature measurement: digital thermometers, differential thermometers, infrared digital thermometers, infrared radiation thermometers, thermal imaging systems.
- Humidity measurement: wet and dry bulb whirling hygrometers, digital humidity indicators.
- Water flow measurement: computerized flow and differential pressure test sets, ultrasonic flow meters, micronics high temperature sensor set.
- Electrical measurement: digital induction ammeters, digital and analogue network testers, microprocessor controlled gaussmeter for measuring magnetic flux density, voltage condition analysers and multimeters, oscilloscope meters, power disturbance monitors, digital power and power factor indicators, electrical energy load analysers, optical reader for gas and electricity meter

readings, portable appliance insulation testers, mechanical/optical tachometers.

- Illumination inspection: illuminance meters calibrated in lux for illuminance measurement and candela/m^2 for measurement of luminance.
- Combustion analysis: portable electronic combustion analysers, continuous sampling gas detectors, pocket gas monitors, hand held combustion analysers for measuring oxygen, carbon monoxide, nitrogen oxide and flue gas temperature.
- Indoor air quality: carbon monoxide/carbon dioxide analysers, flammable gas monitors.
- Recording equipment: electromechanical chart recorders, microprocessor chart recorders, multichannel data printers, portable computers for storing data from chart recorders, loggers, combustion analysers, etc., multichannel data loggers, mechanical/electronic thermohygrographs.

Appendix 6
Cost benefit tables

These tables have been reproduced from the *CIBSE Guide*, section B18 (1970) by permission of the Chartered Institution of Building Services Engineers.

Present value of a single sum

n (years)	Interest (=100r) (%)										
	3	4	5	6	7	8	9	10	12	15	20
1	0.97087	0.96154	0.95238	0.94340	0.93458	0.92593	0.91743	0.90909	0.89286	0.86957	0.83333
2	0.94260	0.92456	0.90703	0.89000	0.87344	0.85734	0.84168	0.82645	0.79719	0.75614	0.69444
3	0.91514	0.88900	0.86384	0.83962	0.81630	0.79383	0.77218	0.75131	0.71178	0.65752	0.57870
4	0.88849	0.85480	0.82270	0.79209	0.76290	0.73503	0.70843	0.68301	0.63552	0.57175	0.48225
5	0.86261	0.82193	0.78353	0.74726	0.71299	0.68058	0.64993	0.62092	0.56743	0.49718	0.40188
6	0.83748	0.79031	0.74622	0.70496	0.66634	0.63017	0.59627	0.56447	0.50663	0.43233	0.33490
7	0.81309	0.75992	0.71068	0.66506	0.62275	0.58349	0.54703	0.51316	0.45235	0.37594	0.27908
8	0.78941	0.73069	0.67684	0.62741	0.58201	0.54027	0.50187	0.46651	0.40388	0.32690	0.23257
9	0.76642	0.70259	0.64461	0.59190	0.54393	0.50025	0.46043	0.42410	0.36061	0.28426	0.19381
10	0.74409	0.67556	0.61391	0.55839	0.50835	0.46319	0.42241	0.38554	0.32197	0.24718	0.16151
11	0.72242	0.64958	0.58468	0.52679	0.47509	0.42888	0.38753	0.35049	0.28748	0.21494	0.13459
12	0.70138	0.62460	0.55684	0.49697	0.44401	0.39711	0.35553	0.31863	0.25668	0.18691	0.11216
13	0.68095	0.60057	0.53032	0.46884	0.41496	0.36770	0.32618	0.28966	0.22917	0.16253	0.09346
14	0.66112	0.57748	0.50507	0.44230	0.38782	0.34046	0.29925	0.26333	0.20462	0.14133	0.07789
15	0.64186	0.55526	0.48102	0.41727	0.36245	0.31524	0.27454	0.23939	0.18270	0.12289	0.06491
16	0.62317	0.53391	0.45811	0.39365	0.33873	0.29189	0.25187	0.21763	0.16312	0.10686	0.05409
17	0.60502	0.51337	0.43630	0.37136	0.31657	0.27027	0.23107	0.19784	0.14564	0.09293	0.04507
18	0.58739	0.49363	0.41552	0.35034	0.29586	0.25025	0.21199	0.17986	0.13004	0.08081	0.03756
19	0.57029	0.47464	0.39573	0.33051	0.27651	0.23171	0.19449	0.16351	0.11611	0.07027	0.03130
20	0.55368	0.45639	0.37689	0.31180	0.25842	0.21455	0.17843	0.14864	0.10367	0.06110	0.02608
25	0.47761	0.37512	0.29530	0.23300	0.18425	0.14602	0.11597	0.09230	0.05882	0.03038	0.01048
30	0.41199	0.30832	0.23138	0.17411	0.13137	0.09938	0.07537	0.05731	0.03338	0.01510	0.00421
35	0.35538	0.25342	0.18129	0.13011	0.09366	0.06763	0.04899	0.03558	0.01894	0.00751	0.00169
40	0.30656	0.20829	0.14205	0.09722	0.06678	0.04603	0.03184	0.02209	0.01075	0.00373	0.00068
45	0.26444	0.17120	0.11130	0.07265	0.04761	0.03133	0.02069	0.01372	0.00610	0.00186	0.00027
50	0.22811	0.14071	0.08720	0.05429	0.03395	0.02132	0.01345	0.00852	0.00346	0.00092	0.00011
55	0.19677	0.11566	0.06833	0.04057	0.02420	0.01451	0.00874	0.00529	0.00196	0.00044	0.00004
60	0.16973	0.09506	0.05354	0.03031	0.01726	0.00988	0.00568	0.00328	0.00111	0.00023	0.00002

Note:
The value of £1 in n years hence, when discounted at interest rate r per annum $= (1 + r)^{-n}$.

Terminal value of a single sum at compound interest

n (years)	Interest (=100r) (%)										
	3	4	5	6	7	8	9	10	12	15	20
1	1.0300	1.0400	1.0500	1.0600	1.0700	1.0800	1.0900	1.1000	1.1200	1.1500	1.2000
2	1.0609	1.0816	1.1025	1.1236	1.1449	1.1664	1.1881	1.2100	1.2544	1.3225	1.4400
3	1.0927	1.1249	1.1576	1.1910	1.2250	1.2597	1.2950	1.3310	1.4049	1.5209	1.7280
4	1.1255	1.1699	1.2155	1.2625	1.3108	1.3605	1.4116	1.4641	1.5735	1.7490	2.0736
5	1.1593	1.2167	1.2763	1.3382	1.4026	1.4693	1.5386	1.6105	1.7623	2.0114	2.4883
6	1.1941	1.2653	1.3401	1.4185	1.5007	1.5869	1.6771	1.7716	1.9738	2.3131	2.9860
7	1.2299	1.3159	1.4071	1.5036	1.6058	1.7138	1.8280	1.9487	2.2107	2.6600	3.5832
8	1.2668	1.3686	1.4775	1.5938	1.7182	1.8509	1.9926	2.1436	2.4760	3.0590	4.2998
9	1.3048	1.4233	1.5513	1.6895	1.8385	1.9990	2.1719	2.3579	2.7731	3.5179	5.1598
10	1.3439	1.4802	1.6289	1.7908	1.9672	2.1589	2.3674	2.5937	3.1058	4.0456	6.1917
11	1.3842	1.5395	1.7103	1.8983	2.1049	2.3316	2.5804	2.8531	3.4785	4.6524	7.4301
12	1.4258	1.6010	1.7959	2.0122	2.2522	2.5182	2.8127	3.1384	3.8960	5.3502	8.9161
13	1.4685	1.6651	1.8856	2.1329	2.4098	2.7196	3.0658	3.4523	4.3635	6.1528	10.699
14	1.5126	1.7317	1.9799	2.2609	2.5785	2.9372	3.3417	3.7975	4.8871	7.0757	12.839
15	1.5580	1.8009	2.0789	2.3966	2.7590	3.1722	3.6425	4.1772	5.4736	8.1371	15.407
16	1.6047	1.8730	2.1829	2.5404	2.9522	3.4259	3.9703	4.5950	6.1304	9.3576	18.488
17	1.6528	1.9479	2.2920	2.6928	3.1588	3.7000	4.3276	5.0545	6.8660	10.761	22.186
18	1.7024	2.0258	2.4066	2.8543	3.3799	3.9960	4.7171	5.5599	7.6900	12.375	26.623
19	1.7535	2.1068	2.5269	3.0256	3.6165	4.3157	5.1417	6.1159	8.6128	14.232	31.948
20	1.8061	2.1911	2.6533	3.2071	3.8697	4.6610	5.6044	6.7275	9.6463	16.367	38.338
25	2.0938	2.6658	3.3864	4.2919	5.4274	6.8485	8.6231	10.835	17.000	32.919	95.396
30	2.4273	3.2434	4.3219	5.7435	7.6123	10.063	13.268	17.449	29.960	66.212	237.38
35	2.8139	3.9461	5.5160	7.6861	10.677	14.785	20.414	28.102	52.800	133.18	590.67
40	3.2620	4.8010	7.0400	10.286	14.974	21.725	31.409	45.259	93.051	267.86	1469.8
45	3.7816	5.8412	8.9850	13.765	21.002	31.920	48.327	72.890	163.99	538.77	3657.3
50	4.3839	7.1067	11.467	18.420	29.457	46.902	74.358	117.39	289.00	1083.7	9100.4
55	5.0821	8.6464	14.636	24.650	41.315	68.914	114.41	189.06	509.32	2179.7	22 644
60	5.8916	10.519	18.679	32.988	57.946	101.26	176.03	304.50	897.59	4384.1	56 346

Note:
The amount to which £1 will increase in n years with interest rate r per annum $= (1 + r)^n$.

Present value of an annuity

n (years)	3	4	5	6	7	Interest (=100r) (%) 8	9	10	12	15	20
1	0.9709	0.9615	0.9524	0.9434	0.9346	0.9259	0.9174	0.9091	0.8929	0.8696	0.8333
2	1.9135	1.8861	1.8594	1.8334	1.8080	1.7833	1.7591	1.7355	1.6901	1.6257	1.5278
3	2.8286	2.7751	2.7232	2.6730	2.6243	2.5771	2.5313	2.4869	2.4018	2.2832	2.1065
4	3.7171	3.6299	3.5460	3.4651	3.3872	3.3121	3.2397	3.1699	3.0373	2.8550	2.5887
5	4.5797	4.4518	4.3295	4.2124	4.1002	3.9927	3.8897	3.7908	3.6048	3.3522	2.9906
6	5.4172	5.2421	5.0757	4.9173	4.7665	4.6229	4.4859	4.3553	4.1114	3.7845	3.3255
7	6.2303	6.0021	5.7864	5.5824	5.3893	5.2064	5.0330	4.8684	4.5638	4.1604	3.6046
8	7.0197	6.7327	6.4632	6.2098	5.9713	5.7466	5.5348	5.3349	4.9676	4.4873	3.8372
9	7.7861	7.4353	7.1078	6.8017	6.5152	6.2469	5.9952	5.7590	5.3282	4.7716	4.0310
10	8.5302	8.1109	7.7217	7.3601	7.0236	6.7101	6.4177	6.1446	5.6502	5.0188	4.1925
11	9.2526	8.7605	8.3064	7.8869	7.4987	7.1390	6.8052	6.4951	5.9377	5.2337	4.3271
12	9.9540	9.3851	8.8633	8.3838	7.9427	7.5361	7.1607	6.8137	6.1944	5.4206	4.4392
13	10.6350	9.9856	9.3936	8.8527	8.3577	7.9038	7.4869	7.1034	6.4235	5.5831	4.5327
14	11.2961	10.5631	9.8986	9.2950	8.7455	8.2442	7.7862	7.3667	6.6282	5.7245	4.6106
15	11.9379	11.1184	10.3797	9.7122	9.1079	8.5595	8.0607	7.6061	6.8109	5.8474	4.6755
16	12.5611	11.6523	10.8378	10.1059	9.4466	8.8514	8.3126	7.8237	6.9740	5.9542	4.7296
17	13.1661	12.1637	11.2741	10.4773	9.7632	9.1216	8.5436	8.0216	7.1196	6.0472	4.7746
18	13.7535	12.6593	11.6896	10.8276	10.0591	9.3719	8.7556	8.2014	7.2497	6.1280	4.8122
19	14.3238	13.1339	12.0853	11.1581	10.3356	9.6036	8.9501	8.3649	7.3658	6.1982	4.8435
20	14.8775	13.5903	12.4622	11.4699	10.5940	9.8181	9.1285	8.5136	7.4694	6.2593	4.8696
25	17.4131	15.6221	14.0939	12.7834	11.6536	10.6748	9.8226	9.0770	7.8431	6.4641	4.9476
30	19.6004	17.2920	15.3725	13.7648	12.4090	11.2578	10.2737	9.4269	8.0552	6.5660	4.9789
35	21.4872	18.6646	16.3742	14.4982	12.9477	11.6546	10.5668	9.6442	8.1755	6.6166	4.9915
40	23.1148	19.7928	17.1591	15.0463	13.3317	11.9246	10.7574	9.7791	8.2438	6.6418	4.9966
45	24.5187	20.7200	17.7741	15.4558	13.6055	12.1084	10.8812	9.8628	8.2825	6.6543	4.9986
50	25.7298	21.4822	18.2559	15.7619	13.8007	12.2335	10.9617	9.9148	8.3045	6.6605	4.9995
55	26.7744	22.1086	18.6335	15.9905	13.9400	12.3186					
60	27.6756	22.6235	18.9293	16.1614	14.0392	12.3766					

Notes:

The present value of £1 per annum for n years when discounted at interest rate r per annum $= [(1 - (1 + r)^{-n})/r]$.

The amount per annum to redeem a loan of £1 at the end of n years and provide interest on the outstanding balance at r per annum can be determined from the reciprocals of values in this table.

Appendix 7
Standard service illuminance for various activities/interiors

This table has been reproduced from the *CIBSE Guide*, Section A1 (1986) by permission of the Chartered Institution of Building Services Engineers.

Standard Service Illuminance (lx)	*Characteristics of the activity/interior*	*Representative activities/interiors*
50	Interiors visited rarely with visual tasks confined to movement and casual seeing without perception of detail.	Cable tunnels, indoor storage tanks, walkways.
100	Interiors visited occasionally with visual tasks confined to movement and casual seeing calling for only limited perception of detail.	Corridors, changing rooms, bulk stores.
150	Interiors visited occasionally with visual tasks requiring some perception of detail or involving some risk to people, plant or product.	Loading bays, medical stores, switchrooms.
200	Continuously occupied interiors, visual tasks not requiring any perception or detail.	Monitoring automatic processes in manufacture casting concrete, turbine halls.
300	Continuously occupied interiors, visual tasks moderately easy, i.e. large details >10 min arc and/or high contrast.	Packing goods, rough core making in foundries, rough sawing.
500	Visual tasks moderately difficult, i.e. details to be seen are of moderate size (5–10 min arc) and may be of low contrast. Also colour judgement may be required.	General offices, engine assembly, painting and spraying.
750	Visual tasks difficult, i.e. details to be seen are small (3–5 min arc) and of low contrast, also good colour judgements may be required.	Drawing offices, ceramic decoration, meat inspection

1000	Visual tasks very difficult, i.e. details to be seen are very small (2–3 min arc) and can be of very low contrast. Also accurate colour adjustments may be required.	Electronic component assembly, gauge and tool rooms, retouching paintwork.
1500	Visual tasks extremely difficult, i.e. details to be seen extremely small (1–2 min arc) and of low contrast. Visual aids may be of advantage.	Inspection of graphic reproduction, hand tailoring, fine die sinking.
2000	Visual tasks exceptionally difficult, i.e. details to be seen exceptionally small (<1 min arc) with very low contrasts. Visual aids will be of advantage.	Assembly of minute mechanisms, finished fabric inspection.

Appendix 8
Source organizations and addresses

BRE: Building Research Establishment, Garston, Watford WD2 7JR.

BRECSU: Building Research Energy Conservation Support Unit, Enquiries Unit, Building Research Establishment, Garston, Watford WD2 7JR.

BSRIA: Building Services Research and Information Association, Old Bracknell Lane, Bracknell, Berkshire RG12 7AH.

CIBSE: Chartered Institution of Building Services Engineers, 222 Balham High Road, London SW12 9BS.

EEO: Energy Efficiency Office, as for BRECSU.
Note the previous title of the Energy Efficiency Office has been discontinued (April 1996) and the work is now promoted under the name of the Energy Efficiency Best Practice Programme of the Department of the Environment.

ETSU: Energy Technology Support Unit, Harwell, Didcot, Oxfordshire OX11 0RA.

HVCA: Heating and Ventilating Contractors Association, ESCA House, 34 Palace Court, Bayswater, London W2 4JG.

The Meteorological Office, Services and Business, Room 124b Main Building, London Road, Bracknell, Berkshire RG12 2SZ.

Appendix 9
Source journals

BSEE: Building Services Environmental Engineer Journal

Building Services, the CIBSE Journal

eibi: Energy in Buildings and Industry Journal

Energy Efficiency Best Practice Programme Guide No. 187, published by the DOE

Energy Management Journal

HAC: Heating and Air Conditioning Journal

Appendix 10
Some current energy saving schemes

BREEAM: Building Research Establishment Environmental Assessment Method.

EMAS: Ecological Management and Audit Scheme; The European Community register for local authorities.

TEWI: Total Equivalent Warming Impact Analysis; Energy consumption of refrigeration plant over its working life.

Bibliography

[1]   *CIBSE Guide*, Section A9, 1986.   **Chapter 1**
[2]   K. J. Moss, *Heating and Water Services Design in Buildings*,
      E&FN Spon, 1996.
[3]   *CIBSE Guide*, Section A3, 1986.
[4]   *CIBSE Guide*, Section A7, 1986.
[5]   *BSRIA Rules of Thumb*, 1995.
[6]   *BSEE Journal*, April 1996.
[7]   Standard Degree Days, *Energy Management Journal*.
[8]   EEO, *Fuel Efficiency Booklet No. 7*, 1993.
[9]   *BSEE Journal*, February 1980.
[10] *Building Services*, the CIBSE Journal, January 1977.
[11] *Building Services*, the CIBSE Journal, July 1980.
[12] *Building Services*, the CIBSE Journal, July 1991.
[13] *Building Services*, the CIBSE Journal, March 1995.

[1]   *BSRIA Rules of Thumb*, 1995.   **Chapter 2**
[2]   Standard Degree Days, *Energy Management Journal*.
[3]   *CIBSE Guide*, Section A3, 1986.
[4]   *CIBSE Guide*, Section B18, 1986.

[1]   *CIBSE Guide*, Section B18, 1986.   **Chapter 3**
[2]   *CIBSE Guide*, Section A7, 1986.
[3]   *Energy Management Journal*, Jan/Feb 1996.

[1]   *CIBSE Guide*, Section B4, 1970.   **Chapter 4**
[2]   *CIBSE Building Energy Code*, part 4, 1982.

[1]   The Meteorological Office, cooling SDDs for the Thames   **Chapter 5**
      Valley, 20-year period to 1995.
[2]   EEO, *Fuel Efficiency Booklet No. 7*, 1993.
[3]   SDD *Energy Management Journal*.

[4] *CIBSE Guide*, Section A6, 1970.
[5] *CIBSE Guide*, Section A9, 1986.
[6] *BSRIA Rules of Thumb*, 1995.
[7] *CIBSE Guide*, Section A7, 1986.
[8] *eibi journal*, February 1996.
[9] *BSEE Journal*, April 1996.

Chapter 6

[1] *CIBSE Building Energy Code*, part 4, 1982.
[2] *Building Services*, the CIBSE Journal, November 1995, December 1995, February 1996, March 1996.
[3] *Energy Management Journal*, July/August 1995.

Chapter 7

[1] *Energy Management Journal*, Jan/Feb 1996.
[2] *CIBSE Building Energy Code*, part 2, 1981.
[3] *CIBSE Guide*, Section B18, 1970.
[4] HVCA open learning package: *Energy Management*, 1989.
[5] *HAC Journal*, June 1985.
[6] D. V. Chadderton, *Air Conditioning. A practical introduction*, E&FN Spon, 1993.
[7] K. J. Moss, *Heating and Water Services Design in Buildings*, E&FN Spon, 1996.

Chapter 8

[1] *CIBSE Guide*, Section B18, 1986.

Chapter 9

[1] EEO, *Fuel Efficiency Booklet No. 1, Energy audits in buildings*, 1995.
[2] *Building Services*, the CIBSE Journal, November 1995, December 1995, February 1996, March 1996.
[3] EEO, *Fuel Efficiency Booklet No. 1, Energy audits in buildings*, 1995; *No. 9, Economic use of electricity in buildings*, 1995; *No. 10, Controls and energy savings*, 1995; *No. 12, Energy management and good lighting practice*, 1993.

Chapter 10

[1] *CIBSE Building Energy Code*, part 4, 1982.
[2] *Energy Management Journal*.

Index

Page numbers in **bold** indicate figures, and page numbers in *italic*, tables.

DATE DUE

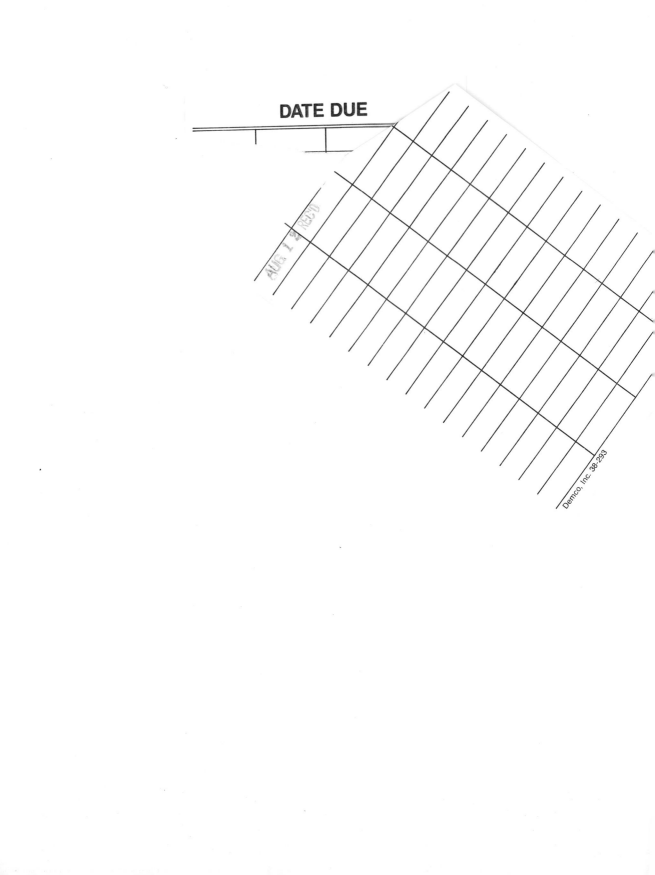

Demco, Inc. 38-293